ALEKS MATH WORKSHEETS 2024

A Comprehensive Review of ALEKS Math Test

By

Reza Nazari

Published by: **Effortless Math Education Inc.**

For Online Math Practice Visit www.EffortlessMath.com

Introdcution

Dive into "**ALEKS Math Worksheets**", a professional-grade resource designed for clarity and effectiveness in preparing for the ALEKS math section. This book is your ally in demystifying math, laid out according to the official guidelines for the ALEKS exam. It presents a broad range of questions and exercises that cover all the necessary topics, ensuring you're well-versed in the kinds of problems you'll encounter on the test.

Accompanying the book is a well-structured online course, creating a comprehensive learning package. Each chapter in the book is connected to an online component through a QR code and a web link. These links direct you to in-depth lessons, worked-out examples, practice exercises, video tutorials, and extra worksheets — all aligned to reinforce the material in the book.

Furthermore, we've included answers to all the exercises, allowing for self-assessment and targeted study. This integration of book and digital resources provides a robust platform for all learners to develop their mathematical skills in a focused and practical manner.

ALEKS Math Worksheets is crafted to offer straightforward, professional guidance for those aiming to excel in their ALEKS test. It's not just about studying; it's about achieving a deep and lasting understanding of math that will serve you well on test day and beyond.

Effortless Math's ALEKS Online Center

Effortless Math Online ALEKS Center offers a complete study program, including the following:

✓ Step-by-step instructions on how to prepare for the ALEKS Math test

✓ Numerous ALEKS Math worksheets to help you measure your math skills

✓ Complete list of ALEKS Math formulas

✓ Video lessons for all ALEKS Math topics

✓ Full-length ALEKS Math practice tests

✓ And much more...

No Registration Required.

Visit **EffortlessMath.com/ALEKS** to find your online ALEKS Math resources.

How to Use This Book Effectively

L ook no further when you need a practice book to improve your math skills to succeed on the math portion of the ALEKS test. Each section of this comprehensive practice book will provide you with the knowledge, tools, and understanding needed to succeed on the test.

It's imperative that you understand each practice question before moving onto another one, as that's the way to guarantee your success. Each practice test provides you with a step-by-step guide of every question to better understand the content that will be on the test. To get the best possible results from this book:

➢ **Begin studying long before your test date.** This provides you ample time to learn the different math concepts. The earlier you begin studying for the test, the sharper your skills will be. Do not procrastinate! Provide yourself with plenty of time to learn the concepts and feel comfortable that you understand them when your test date arrives.

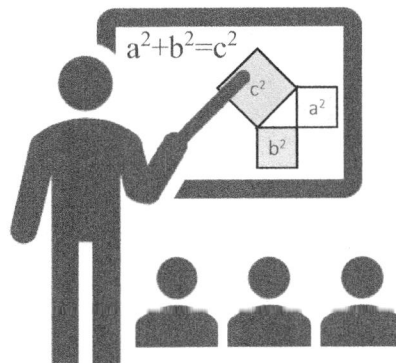

➢ **Practice consistently.** Study ALEKS Math concepts at least 20 to 30 minutes a day. Remember, slow and steady wins the race, which can be applied to preparing for the ALEKS Math test. Instead of cramming to tackle everything at once, be patient and learn the math topics in short bursts.

➢ Whenever you get a math problem wrong, **mark it off, and review it later** to make sure you understand the concept.

➢ Once you've reviewed the book's instructions, **take a practice test** to gauge your level of readiness. Then, review your results. Read detailed answers and solutions for each question you missed.

➢ **Take another practice test** to get an idea of how ready you are to take the actual exam. Taking the practice tests will give you the confidence you need on test day. Simulate the ALEKS testing environment by sitting in a quiet room free from distraction. Make sure to clock yourself with a timer.

How to Study for the ALEKS Math Test

Studying for the ALEKS Math test can be a really daunting and boring task. What's the best way to go about it? Is there a certain study method that works better than others? Well, studying for the ALEKS Math can be done effectively. The following six-step program has been designed to make preparing for the ALEKS Math test more efficient and less overwhelming.

Step **1** - Create a study plan
Step **2** - Choose your study resources
Step **3** - Review, Learn, Practice
Step **4** - Learn and practice test-taking strategies
Step **5** - Learn the ALEKS Test format and take practice tests
Step **6** - Analyze your performance

STEP 1: Create a Study Plan

It's always easier to get things done when you have a plan. Creating a study plan for the ALEKS Math test can help you to stay on track with your studies. It's important to sit down and prepare a study plan with what works with your life, work, and any other obligations you may have. Devote enough time each day to studying. It's also a great idea to break down each section of the exam into blocks and study one concept at a time.

It's important to understand that there is no "right" way to create a study plan. Your study plan will be personalized based on your specific needs and learning style.

Follow these guidelines to create an effective study plan for your ALEKS Math test:

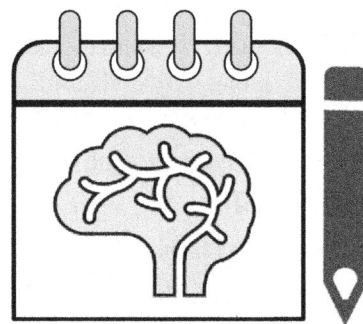

★ **Analyze your learning style and study habits** – Everyone has a different learning style. It is essential to embrace your individuality and the unique way you learn. Think about what works and what doesn't work for you. Do you prefer ALEKS Math prep books or a combination of textbooks and video lessons? Does it work better for you if you study every night for thirty minutes or is it more effective to study in the morning before going to work?

★ **Evaluate your schedule** – Review your current schedule and find out how much time you can consistently devote to ALEKS Math study.

★ **Develop a schedule** – Now it's time to add your study schedule to your calendar like any other obligation. Schedule time for study, practice, and review. Plan out which topic you will study on which day to ensure that you're devoting enough time to each concept. Develop a study plan that is mindful, realistic, and flexible.

★ **Stick to your schedule** – A study plan is only effective when it is followed consistently. You should try to develop a study plan that you can follow for the length of your study program.

★ **Evaluate your study plan and adjust as needed** – Sometimes you need to adjust your plan when you have new commitments. Check in with yourself regularly to make sure that you're not falling behind in your study plan. Remember, the most important thing is sticking to your plan. Your study plan is all about helping you be more productive. If you find that your study plan is not as effective as you want, don't get discouraged. It's okay to make changes as you figure out what works best for you.

STEP 2: Choose Your Study Resources

There are numerous textbooks and online resources available for the ALEKS Math test, and it may not be clear where to begin. Don't worry! Effortless Math's ALEKS online center provides everything you need to fully prepare for your ALEKS Math test. In addition to the practice tests in this book, you can also use Effortless Math's online resources. (video lessons, worksheets, formulas, etc.)

Simply visit EffortlessMath.com/ALEKS to find your online ALEKS Math resources.

STEP 3: Review, Learn, Practice

Effortless Math's ALEKS course breaks down each subject into specific skills or content areas. For instance, the percent concept is separated into different topics–percent calculation, percent increase and decrease, percent problems, etc. Use our online resources to help you go over all key math concepts and topics on the ALEKS Math test.

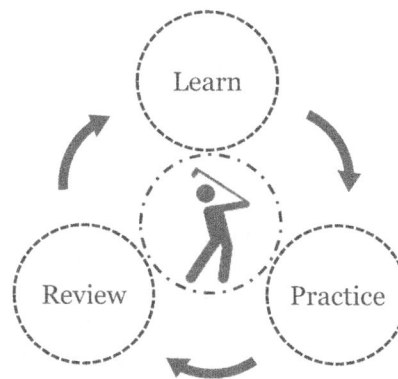

As you review each concept, take notes or highlight the concepts you would like to go over again in the future. If you're unfamiliar with a topic or something is difficult for you, do additional research on it. For each math topic, plenty of instructions, step-by-step guides, and examples are provided to ensure you get a good grasp of the material. You can also find video lessons on the Effortless Math website for each ALEKS Math concept.

Quickly review the topics you do understand to get a brush-up of the material. Be sure to use the worksheets and do the practice questions provided on the Effortless Math's online center to measure your understanding of the concepts.

STEP 4: Learn and Practice Test-taking Strategies

In the following sections, you will find important test-taking strategies and tips that can help you earn extra points. You'll learn how to think strategically and when to guess if you don't know the answer to a question. Using ALEKS Math test-taking strategies and tips can help you raise your score and do well on the test. Apply test taking strategies on the practice tests to help you boost your confidence.

STEP 5: Learn the ALEKS Test Format and Take Practice Tests

The ALEKS *Test Review* section provides information about the structure of the ALEKS test. Read this section to learn more about the ALEKS test structure, different test sections, the number of questions in each section, and the section time limits. When you have a prior understanding of the test format and different types of ALEKS Math questions, you'll feel more confident when you take the actual exam.

Once you have read through the instructions and lessons and feel like you are ready to go – take advantage of the full-length ALEKS Math practice tests available in this book. Use the practice tests to sharpen your skills and build confidence.

The ALEKS Math practice tests offered in the book are formatted similarly to the actual ALEKS Math test. When you take each practice test, try to simulate actual testing conditions. To take the practice tests, sit in a quiet space, time yourself, and work through as many of the questions as time allows. The practice tests are followed by detailed answer explanations to help you find your weak areas, learn from your mistakes, and raise your ALEKS Math score.

STEP 6: Analyze Your Performance

After taking the practice tests, look over the answer keys and explanations to learn which questions you answered correctly and which you did not. Never be discouraged if you make a few mistakes. See them as a learning opportunity. This will highlight your strengths and weaknesses.

You can use the results to determine if you need additional practice or if you are ready to take the actual ALEKS Math test.

Looking for more?

Visit EffortlessMath.com/ALEKS to find hundreds of ALEKS Math worksheets, video tutorials, practice tests, ALEKS Math formulas, and much more.

Or scan this QR code.

No Registration Required.

ALEKS Test Review

ALEKS (Assessment and Learning in Knowledge Spaces) is an artificial intelligence-based assessment tool to measure the strengths and weaknesses of a student's mathematical knowledge. ALEKS is available for a variety of subjects and courses in K-12, Higher Education, and Continuing Education. The findings of ALEKS's assessment test help to find an appropriate level for course placement. The ALEKS math placement assessment ensures students' readiness for particular math courses at colleges.

ALEKS does not use multiple-choice questions like most other standardized tests. Instead, it utilizes adaptable and easy-to-use method that mimic paper and pencil techniques. When taking the ALEKS test, a brief tutorial helps you learn how to use ALEKS answer input tools. You then begin the ALEKS Assessment. In about 30 to 45 minutes, the test measures your current content knowledge by asking 20 to 30 questions. ALEKS is a Computer Adaptive (CA) assessment. It means that each question will be chosen on the basis of answers to all the previous questions. Therefore, each set of assessment questions is unique. The ALEKS Math assessment does not allow you to use a personal calculator. But for some questions ALEKS onscreen calculator button is active and the test taker can use it.

Key Features of the ALEKS Mathematics Assessment

Some key features of the ALEKS Math assessment are:

- ❖ Mathematics questions on ALEKS are adaptive to identify the student's knowledge from a comprehensive standard curriculum, ranging from basic arithmetic up to precalculus, including trigonometry but not calculus.

- ❖ Unlike other standardized tests, the ALEKS assessment does not provide a "grade" or "raw score." Instead, ALEKS identifies which concepts the student has mastered and what topics the student needs to learn.

- ❖ ALEKS does not use multiple-choice questions. Instead, students need to produce authentic mathematical input.

- ❖ There is no time limit for taking the ALEKS Math assessment. But it usually takes 30 to 45 minutes to complete the assessment.

The ALEKS Math score is between 1 and 100 and is interpreted as a percentage correct. A higher ALEKS score indicates that the test-taker has mastered more math concepts. ALEKS Math assessment tool evaluates mastery of a comprehensive set of mathematics skills ranging from basic arithmetic up to precalculus, including trigonometry but not calculus. It will place students in classes up to Calculus.

Contents

Simplifying Fractions

✎ *Simplify each fraction.*

1) $\frac{10}{15} =$

2) $\frac{8}{20} =$

3) $\frac{12}{42} =$

4) $\frac{5}{20} =$

5) $\frac{6}{18} =$

6) $\frac{18}{27} =$

7) $\frac{15}{55} =$

8) $\frac{24}{54} =$

9) $\frac{63}{72} =$

10) $\frac{40}{64} =$

11) $\frac{23}{46} =$

12) $\frac{35}{63} =$

13) $\frac{32}{36} =$

14) $\frac{81}{99} =$

15) $\frac{16}{64} =$

16) $\frac{14}{35} =$

17) $\frac{19}{38} =$

18) $\frac{18}{54} =$

19) $\frac{56}{70} =$

20) $\frac{40}{45} =$

21) $\frac{9}{90} =$

22) $\frac{20}{25} =$

23) $\frac{32}{48} =$

24) $\frac{7}{49} =$

25) $\frac{18}{48} =$

26) $\frac{54}{108} =$

Simplifying Fractions - Answers

 *Simplify each fraction.*

1) $\dfrac{10}{15} = \dfrac{2}{3}$

2) $\dfrac{8}{20} = \dfrac{2}{5}$

3) $\dfrac{12}{42} = \dfrac{2}{7}$

4) $\dfrac{5}{20} = \dfrac{1}{4}$

5) $\dfrac{6}{18} = \dfrac{1}{3}$

6) $\dfrac{18}{27} = \dfrac{2}{3}$

7) $\dfrac{15}{55} = \dfrac{3}{11}$

8) $\dfrac{24}{54} = \dfrac{4}{9}$

9) $\dfrac{63}{72} = \dfrac{7}{8}$

10) $\dfrac{40}{64} = \dfrac{5}{8}$

11) $\dfrac{23}{46} = \dfrac{1}{2}$

12) $\dfrac{35}{63} = \dfrac{5}{9}$

13) $\dfrac{32}{36} = \dfrac{8}{9}$

14) $\dfrac{81}{99} = \dfrac{9}{11}$

15) $\dfrac{16}{64} = \dfrac{1}{4}$

16) $\dfrac{14}{35} = \dfrac{2}{5}$

17) $\dfrac{19}{38} = \dfrac{1}{2}$

18) $\dfrac{18}{54} = \dfrac{1}{3}$

19) $\dfrac{56}{70} = \dfrac{4}{5}$

20) $\dfrac{40}{45} = \dfrac{8}{9}$

21) $\dfrac{9}{90} = \dfrac{1}{10}$

22) $\dfrac{20}{25} = \dfrac{4}{5}$

23) $\dfrac{32}{48} = \dfrac{2}{3}$

24) $\dfrac{7}{49} = \dfrac{1}{7}$

25) $\dfrac{18}{48} = \dfrac{3}{8}$

26) $\dfrac{54}{108} = \dfrac{1}{2}$

Adding and Subtracting Fractions

✎ *Calculate and write the answer in the lowest term.*

1) $\dfrac{1}{5} + \dfrac{1}{7} =$

2) $\dfrac{3}{7} + \dfrac{4}{5} =$

3) $\dfrac{3}{8} - \dfrac{1}{9} =$

4) $\dfrac{4}{5} - \dfrac{5}{9} =$

5) $\dfrac{2}{9} + \dfrac{1}{3} =$

6) $\dfrac{3}{10} + \dfrac{2}{5} =$

7) $\dfrac{9}{10} - \dfrac{4}{5} =$

8) $\dfrac{7}{9} - \dfrac{3}{7} =$

9) $\dfrac{3}{4} + \dfrac{1}{3} =$

10) $\dfrac{3}{8} + \dfrac{2}{5} =$

11) $\dfrac{3}{4} - \dfrac{2}{5} =$

12) $\dfrac{7}{9} - \dfrac{2}{3} =$

13) $\dfrac{4}{9} + \dfrac{5}{6} =$

14) $\dfrac{2}{3} + \dfrac{1}{4} =$

15) $\dfrac{9}{10} - \dfrac{3}{5} =$

16) $\dfrac{7}{12} - \dfrac{1}{2} =$

17) $\dfrac{4}{5} + \dfrac{2}{3} =$

18) $\dfrac{5}{7} + \dfrac{1}{5} =$

19) $\dfrac{5}{9} - \dfrac{2}{5} =$

20) $\dfrac{3}{5} - \dfrac{2}{9} =$

21) $\dfrac{7}{9} + \dfrac{1}{7} =$

22) $\dfrac{5}{8} + \dfrac{2}{3} =$

23) $\dfrac{4}{7} + \dfrac{2}{3} =$

24) $\dfrac{6}{7} - \dfrac{4}{9} =$

25) $\dfrac{4}{5} - \dfrac{2}{15} =$

26) $\dfrac{2}{9} + \dfrac{4}{5} =$

Adding and Subtracting Fractions - Answers

Calculate and write the answer in the lowest term.

1) $\frac{1}{5} + \frac{1}{7} = \frac{12}{35}$

2) $\frac{3}{7} + \frac{4}{5} = \frac{43}{35}$

3) $\frac{3}{8} - \frac{1}{9} = \frac{19}{72}$

4) $\frac{4}{5} - \frac{5}{9} = \frac{11}{45}$

5) $\frac{2}{9} + \frac{1}{3} = \frac{5}{9}$

6) $\frac{3}{10} + \frac{2}{5} = \frac{7}{10}$

7) $\frac{9}{10} - \frac{4}{5} = \frac{1}{10}$

8) $\frac{7}{9} - \frac{3}{7} = \frac{22}{63}$

9) $\frac{3}{4} + \frac{1}{3} = \frac{13}{12}$

10) $\frac{3}{8} + \frac{2}{5} = \frac{31}{40}$

11) $\frac{3}{4} - \frac{2}{5} = \frac{7}{20}$

12) $\frac{7}{9} - \frac{2}{3} = \frac{1}{9}$

13) $\frac{4}{9} + \frac{5}{6} = \frac{23}{18}$

14) $\frac{2}{3} + \frac{1}{4} = \frac{11}{12}$

15) $\frac{9}{10} - \frac{3}{5} = \frac{3}{10}$

16) $\frac{7}{12} - \frac{1}{2} = \frac{1}{12}$

17) $\frac{4}{5} + \frac{2}{3} = \frac{22}{15}$

18) $\frac{5}{7} + \frac{1}{5} = \frac{32}{35}$

19) $\frac{5}{9} - \frac{2}{5} = \frac{7}{45}$

20) $\frac{3}{5} - \frac{2}{9} = \frac{17}{45}$

21) $\frac{7}{9} + \frac{1}{7} = \frac{58}{63}$

22) $\frac{5}{8} + \frac{2}{3} = \frac{31}{24}$

23) $\frac{4}{7} + \frac{2}{3} = \frac{26}{21}$

24) $\frac{6}{7} - \frac{4}{9} = \frac{26}{63}$

25) $\frac{4}{5} - \frac{2}{15} = \frac{2}{3}$

26) $\frac{2}{9} + \frac{4}{5} = \frac{46}{45}$

Multiplying and dividing fraction

✏️ *Solve and write the answer in lowest term.*

1) $\frac{1}{2} \times \frac{4}{5} =$

2) $\frac{1}{5} \times \frac{6}{7} =$

3) $\frac{1}{3} \div \frac{1}{7} =$

4) $\frac{1}{7} \div \frac{3}{8} =$

5) $\frac{2}{3} \times \frac{4}{7} =$

6) $\frac{5}{7} \times \frac{3}{4} =$

7) $\frac{2}{5} \div \frac{3}{7} =$

8) $\frac{3}{7} \div \frac{5}{8} =$

9) $\frac{3}{8} \times \frac{4}{7} =$

10) $\frac{2}{9} \times \frac{6}{11} =$

11) $\frac{1}{10} \div \frac{3}{8} =$

12) $\frac{3}{10} \div \frac{4}{5} =$

13) $\frac{6}{7} \times \frac{4}{9} =$

14) $\frac{3}{7} \times \frac{5}{6} =$

15) $\frac{7}{9} \div \frac{6}{11} =$

16) $\frac{1}{15} \div \frac{2}{3} =$

17) $\frac{1}{13} \times \frac{1}{2} =$

18) $\frac{1}{12} \times \frac{4}{7} =$

19) $\frac{1}{15} \div \frac{4}{9} =$

20) $\frac{1}{16} \div \frac{1}{2} =$

21) $\frac{4}{7} \times \frac{5}{8} =$

22) $\frac{1}{11} \times \frac{4}{5} =$

23) $\frac{1}{18} \div \frac{5}{6} =$

24) $\frac{1}{15} \div \frac{3}{8} =$

25) $\frac{1}{11} \times \frac{3}{4} =$

26) $\frac{1}{14} \times \frac{2}{3} =$

bit.ly/3haSiQ — Find more at

Multiplying and Dividing Fractions - Answers

✎ *Solve and write the answer in lowest terms.*

1) $\dfrac{1}{2} \times \dfrac{4}{5} = \dfrac{2}{5}$

2) $\dfrac{1}{5} \times \dfrac{6}{7} = \dfrac{6}{35}$

3) $\dfrac{1}{3} \div \dfrac{1}{7} = \dfrac{7}{3}$

4) $\dfrac{1}{7} \div \dfrac{3}{8} = \dfrac{8}{21}$

5) $\dfrac{2}{3} \times \dfrac{4}{7} = \dfrac{8}{21}$

6) $\dfrac{5}{7} \times \dfrac{3}{4} = \dfrac{15}{28}$

7) $\dfrac{2}{5} \div \dfrac{3}{7} = \dfrac{14}{15}$

8) $\dfrac{3}{7} \div \dfrac{5}{8} = \dfrac{24}{35}$

9) $\dfrac{3}{8} \times \dfrac{4}{7} = \dfrac{3}{14}$

10) $\dfrac{2}{9} \times \dfrac{6}{11} = \dfrac{4}{33}$

11) $\dfrac{1}{10} \div \dfrac{3}{8} = \dfrac{4}{15}$

12) $\dfrac{3}{10} \div \dfrac{4}{5} = \dfrac{3}{8}$

13) $\dfrac{6}{7} \times \dfrac{4}{9} = \dfrac{8}{21}$

14) $\dfrac{3}{7} \times \dfrac{5}{6} = \dfrac{5}{14}$

15) $\dfrac{7}{9} \div \dfrac{6}{11} = \dfrac{77}{54}$

16) $\dfrac{1}{15} \div \dfrac{2}{3} = \dfrac{1}{10}$

17) $\dfrac{1}{13} \times \dfrac{1}{2} = \dfrac{1}{26}$

18) $\dfrac{1}{12} \times \dfrac{4}{7} = \dfrac{1}{21}$

19) $\dfrac{1}{15} \div \dfrac{4}{9} = \dfrac{3}{20}$

20) $\dfrac{1}{16} \div \dfrac{1}{2} = \dfrac{1}{8}$

21) $\dfrac{4}{7} \times \dfrac{5}{8} = \dfrac{5}{14}$

22) $\dfrac{1}{11} \times \dfrac{4}{5} = \dfrac{4}{55}$

23) $\dfrac{1}{18} \div \dfrac{5}{6} = \dfrac{1}{15}$

24) $\dfrac{1}{15} \div \dfrac{3}{8} = \dfrac{8}{45}$

25) $\dfrac{1}{11} \times \dfrac{3}{4} = \dfrac{3}{44}$

26) $\dfrac{1}{14} \times \dfrac{2}{3} = \dfrac{1}{21}$

Adding Mixed Numbers

✍ *Solve and write the answer in lowest terms.*

1) $3\frac{1}{5} + 2\frac{2}{9} =$

2) $1\frac{1}{7} + 5\frac{2}{5} =$

3) $4\frac{4}{5} + 1\frac{2}{7} =$

4) $2\frac{4}{7} + 2\frac{3}{5} =$

5) $1\frac{5}{6} + 1\frac{2}{5} =$

6) $3\frac{5}{7} + 1\frac{2}{9} =$

7) $3\frac{5}{8} + 2\frac{1}{3} =$

8) $1\frac{6}{7} + 3\frac{2}{9} =$

9) $2\frac{5}{9} + 1\frac{1}{4} =$

10) $3\frac{7}{9} + 2\frac{5}{6} =$

11) $2\frac{1}{10} + 2\frac{2}{5} =$

12) $1\frac{3}{10} + 3\frac{4}{5} =$

13) $3\frac{1}{12} + 2\frac{1}{3} =$

14) $5\frac{1}{11} + 1\frac{1}{2} =$

15) $3\frac{1}{21} + 2\frac{2}{3} =$

16) $4\frac{1}{24} + 1\frac{5}{8} =$

17) $2\frac{1}{25} + 3\frac{3}{5} =$

18) $3\frac{1}{15} + 2\frac{2}{10} =$

19) $5\frac{6}{7} + 2\frac{1}{3} =$

20) $2\frac{1}{8} + 3\frac{3}{4} =$

21) $2\frac{5}{7} + 2\frac{2}{21} =$

22) $4\frac{1}{6} + 1\frac{4}{5} =$

23) $3\frac{5}{6} + 1\frac{2}{7} =$

24) $2\frac{7}{8} + 3\frac{1}{3} =$

25) $3\frac{1}{17} + 1\frac{1}{2} =$

26) $1\frac{1}{18} + 1\frac{4}{9} =$

Adding Mixed Numbers - Answers

✍ *Solve and write the answer in lowest terms.*

1) $3\frac{1}{5} + 2\frac{2}{9} = 5\frac{19}{45}$

2) $1\frac{1}{7} + 5\frac{2}{5} = 6\frac{19}{35}$

3) $4\frac{4}{5} + 1\frac{2}{7} = 6\frac{3}{35}$

4) $2\frac{4}{7} + 2\frac{3}{5} = 5\frac{6}{35}$

5) $1\frac{5}{6} + 1\frac{2}{5} = 3\frac{7}{30}$

6) $3\frac{5}{7} + 1\frac{2}{9} = 4\frac{59}{63}$

7) $3\frac{5}{8} + 2\frac{1}{3} = 5\frac{23}{24}$

8) $1\frac{6}{7} + 3\frac{2}{9} = 5\frac{5}{63}$

9) $2\frac{5}{9} + 1\frac{1}{4} = 3\frac{29}{36}$

10) $3\frac{7}{9} + 2\frac{5}{6} = 6\frac{11}{18}$

11) $2\frac{1}{10} + 2\frac{2}{5} = 4\frac{1}{2}$

12) $1\frac{3}{10} + 3\frac{4}{5} = 5\frac{1}{10}$

13) $3\frac{1}{12} + 2\frac{1}{3} = 5\frac{5}{12}$

14) $5\frac{1}{11} + 1\frac{1}{2} = 6\frac{13}{22}$

15) $3\frac{1}{21} + 2\frac{2}{3} = 5\frac{5}{7}$

16) $4\frac{1}{24} + 1\frac{5}{8} = 5\frac{2}{3}$

17) $2\frac{1}{25} + 3\frac{3}{5} = 5\frac{16}{25}$

18) $3\frac{1}{15} + 2\frac{2}{10} = 5\frac{4}{15}$

19) $5\frac{6}{7} + 2\frac{1}{3} = 8\frac{4}{21}$

20) $2\frac{1}{8} + 3\frac{3}{4} = 5\frac{7}{8}$

21) $2\frac{5}{7} + 2\frac{2}{21} = 4\frac{17}{21}$

22) $4\frac{1}{6} + 1\frac{4}{5} = 5\frac{29}{30}$

23) $3\frac{5}{6} + 1\frac{2}{7} = 5\frac{5}{42}$

24) $2\frac{7}{8} + 3\frac{1}{3} = 6\frac{5}{24}$

25) $3\frac{1}{17} + 1\frac{1}{2} = 4\frac{19}{34}$

26) $1\frac{1}{18} + 1\frac{4}{9} = 2\frac{1}{2}$

EffortlessMath.com

Subtracting Mixed Numbers

✎ *Solve and write the answer in lowest terms.*

1) $3\frac{2}{5} - 1\frac{2}{9} =$

2) $5\frac{3}{5} - 1\frac{1}{7} =$

3) $4\frac{2}{5} - 2\frac{2}{7} =$

4) $8\frac{3}{4} - 2\frac{1}{8} =$

5) $9\frac{5}{7} - 7\frac{4}{21} =$

6) $11\frac{7}{12} - 9\frac{5}{6} =$

7) $9\frac{5}{9} - 8\frac{1}{8} =$

8) $13\frac{7}{9} - 11\frac{3}{7} =$

9) $8\frac{7}{12} - 7\frac{3}{8} =$

10) $11\frac{5}{9} - 9\frac{1}{4} =$

11) $6\frac{5}{6} - 2\frac{2}{9} =$

12) $5\frac{7}{8} - 4\frac{1}{3} =$

13) $9\frac{5}{8} - 8\frac{1}{2} =$

14) $4\frac{9}{16} - 2\frac{1}{4} =$

15) $3\frac{2}{3} - 1\frac{2}{15} =$

16) $5\frac{1}{2} - 4\frac{2}{17} =$

17) $5\frac{6}{7} - 2\frac{1}{3} =$

18) $3\frac{3}{7} - 2\frac{2}{21} =$

19) $7\frac{3}{10} - 5\frac{2}{15} =$

20) $4\frac{5}{6} - 2\frac{2}{9} =$

21) $6\frac{3}{7} - 2\frac{2}{9} =$

22) $7\frac{4}{5} - 6\frac{3}{7} =$

23) $10\frac{2}{3} - 9\frac{5}{8} =$

24) $9\frac{3}{4} - 7\frac{4}{9} =$

25) $15\frac{4}{5} - 13\frac{12}{25} =$

26) $13\frac{5}{12} - 7\frac{5}{24} =$

Subtracting Mixed Numbers - Answers

✍ *Solve and write the answer in lowest terms.*

1) $3\frac{2}{5} - 1\frac{2}{9} = 2\frac{8}{45}$

2) $5\frac{3}{5} - 1\frac{1}{7} = 4\frac{16}{35}$

3) $4\frac{2}{5} - 2\frac{2}{7} = 2\frac{4}{35}$

4) $8\frac{3}{4} - 2\frac{1}{8} = 6\frac{5}{8}$

5) $9\frac{5}{7} - 7\frac{4}{21} = 2\frac{11}{21}$

6) $11\frac{7}{12} - 9\frac{5}{6} = 1\frac{3}{4}$

7) $9\frac{5}{9} - 8\frac{1}{8} = 1\frac{31}{72}$

8) $13\frac{7}{9} - 11\frac{3}{7} = 2\frac{22}{63}$

9) $8\frac{7}{12} - 7\frac{3}{8} = 1\frac{5}{24}$

10) $11\frac{5}{9} - 9\frac{1}{4} = 2\frac{11}{36}$

11) $6\frac{5}{6} - 2\frac{2}{9} = 4\frac{11}{18}$

12) $5\frac{7}{8} - 4\frac{1}{3} = 1\frac{13}{24}$

13) $9\frac{5}{8} - 8\frac{1}{2} = 1\frac{1}{8}$

14) $4\frac{9}{16} - 2\frac{1}{4} = 2\frac{5}{16}$

15) $3\frac{2}{3} - 1\frac{2}{15} = 2\frac{8}{15}$

16) $5\frac{1}{2} - 4\frac{2}{17} = 1\frac{13}{34}$

17) $5\frac{6}{7} - 2\frac{1}{3} = 3\frac{11}{21}$

18) $3\frac{3}{7} - 2\frac{2}{21} = 1\frac{1}{3}$

19) $7\frac{3}{10} - 5\frac{2}{15} = 2\frac{1}{6}$

20) $4\frac{5}{6} - 2\frac{2}{9} = 2\frac{11}{18}$

21) $6\frac{3}{7} - 2\frac{2}{9} = 4\frac{13}{63}$

22) $7\frac{4}{5} - 6\frac{3}{7} = 1\frac{13}{35}$

23) $10\frac{2}{3} - 9\frac{5}{8} = 1\frac{1}{24}$

24) $9\frac{3}{4} - 7\frac{4}{9} = 2\frac{11}{36}$

25) $15\frac{4}{5} - 13\frac{12}{25} = 2\frac{8}{25}$

26) $13\frac{5}{12} - 7\frac{5}{24} = 6\frac{5}{24}$

Multiplying Mixed Numbers

✍ *Solve and write the answer in lowest terms.*

1) $1\frac{1}{8} \times 1\frac{3}{4} =$

2) $3\frac{1}{5} \times 2\frac{2}{7} =$

3) $2\frac{1}{8} \times 1\frac{2}{9} =$

4) $2\frac{3}{8} \times 2\frac{2}{5} =$

5) $1\frac{1}{2} \times 5\frac{2}{3} =$

6) $3\frac{1}{2} \times 6\frac{2}{3} =$

7) $9\frac{1}{2} \times 2\frac{1}{6} =$

8) $2\frac{5}{8} \times 8\frac{3}{5} =$

9) $3\frac{4}{5} \times 4\frac{2}{3} =$

10) $5\frac{1}{3} \times 2\frac{2}{7} =$

11) $6\frac{1}{3} \times 3\frac{3}{4} =$

12) $7\frac{2}{3} \times 1\frac{8}{9} =$

13) $8\frac{1}{2} \times 2\frac{1}{6} =$

14) $4\frac{1}{5} \times 8\frac{2}{3} =$

15) $3\frac{1}{8} \times 5\frac{2}{3} =$

16) $2\frac{2}{7} \times 6\frac{2}{5} =$

17) $2\frac{3}{8} \times 7\frac{2}{3} =$

18) $1\frac{7}{8} \times 8\frac{2}{3} =$

19) $9\frac{1}{2} \times 3\frac{1}{5} =$

20) $2\frac{5}{8} \times 4\frac{1}{3} =$

21) $6\frac{1}{3} \times 3\frac{2}{5} =$

22) $5\frac{3}{4} \times 2\frac{2}{7} =$

23) $9\frac{1}{4} \times 2\frac{1}{3} =$

24) $3\frac{3}{7} \times 7\frac{2}{5} =$

25) $4\frac{1}{4} \times 3\frac{2}{5} =$

26) $7\frac{2}{3} \times 3\frac{2}{5} =$

Multiplying Mixed Numbers – Answers

✍ *Solve and write the answer in lowest terms.*

1) $1\frac{1}{8} \times 1\frac{3}{4} = 1\frac{31}{32}$

2) $3\frac{1}{5} \times 2\frac{2}{7} = 7\frac{11}{35}$

3) $2\frac{1}{8} \times 1\frac{2}{9} = 2\frac{43}{72}$

4) $2\frac{3}{8} \times 2\frac{2}{5} = 5\frac{7}{10}$

5) $1\frac{1}{2} \times 5\frac{2}{3} = 8\frac{1}{2}$

6) $3\frac{1}{2} \times 6\frac{2}{3} = 23\frac{1}{3}$

7) $9\frac{1}{2} \times 2\frac{1}{6} = 20\frac{7}{12}$

8) $2\frac{5}{8} \times 8\frac{3}{5} = 22\frac{23}{40}$

9) $3\frac{4}{5} \times 4\frac{2}{3} = 17\frac{11}{15}$

10) $5\frac{1}{3} \times 2\frac{2}{7} = 12\frac{4}{21}$

11) $6\frac{1}{3} \times 3\frac{3}{4} = 23\frac{3}{4}$

12) $7\frac{2}{3} \times 1\frac{8}{9} = 14\frac{13}{27}$

13) $8\frac{1}{2} \times 2\frac{1}{6} = 18\frac{5}{12}$

14) $4\frac{1}{5} \times 8\frac{2}{3} = 36\frac{2}{5}$

15) $3\frac{1}{8} \times 5\frac{2}{3} = 17\frac{17}{24}$

16) $2\frac{2}{7} \times 6\frac{2}{5} = 14\frac{22}{35}$

17) $2\frac{3}{8} \times 7\frac{2}{3} = 18\frac{5}{24}$

18) $1\frac{7}{8} \times 8\frac{2}{3} = 16\frac{1}{4}$

19) $9\frac{1}{2} \times 3\frac{1}{5} = 30\frac{2}{5}$

20) $2\frac{5}{8} \times 4\frac{1}{3} = 11\frac{3}{8}$

21) $6\frac{1}{3} \times 3\frac{2}{5} = 21\frac{8}{15}$

22) $5\frac{3}{4} \times 2\frac{2}{7} = 13\frac{1}{7}$

23) $9\frac{1}{4} \times 2\frac{1}{3} = 21\frac{7}{12}$

24) $3\frac{3}{7} \times 7\frac{2}{5} = 25\frac{13}{35}$

25) $4\frac{1}{4} \times 3\frac{2}{5} = 14\frac{9}{20}$

26) $7\frac{2}{3} \times 3\frac{2}{5} = 26\frac{1}{15}$

Dividing Mixed Numbers

✎ *Solve and write the answer in lowest terms.*

1) $9\frac{1}{2} \div 2\frac{3}{5} =$

2) $2\frac{3}{8} \div 1\frac{2}{5} =$

3) $5\frac{3}{4} \div 2\frac{2}{7} =$

4) $8\frac{1}{3} \div 4\frac{1}{4} =$

5) $7\frac{2}{5} \div 3\frac{3}{4} =$

6) $2\frac{4}{5} \div 3\frac{2}{3} =$

7) $8\frac{3}{5} \div 4\frac{3}{4} =$

8) $6\frac{3}{4} \div 2\frac{2}{9} =$

9) $5\frac{2}{7} \div 2\frac{2}{9} =$

10) $2\frac{2}{5} \div 3\frac{3}{5} =$

11) $4\frac{3}{7} \div 1\frac{7}{8} =$

12) $2\frac{5}{7} \div 2\frac{4}{5} =$

13) $8\frac{3}{5} \div 6\frac{1}{5} =$

14) $2\frac{5}{8} \div 1\frac{8}{9} =$

15) $5\frac{6}{7} \div 2\frac{3}{4} =$

16) $1\frac{3}{5} \div 2\frac{3}{8} =$

17) $5\frac{3}{4} \div 3\frac{2}{5} =$

18) $2\frac{3}{4} \div 3\frac{1}{5} =$

19) $3\frac{2}{3} \div 1\frac{2}{5} =$

20) $4\frac{1}{4} \div 2\frac{2}{3} =$

21) $3\frac{5}{6} \div 2\frac{4}{5} =$

22) $2\frac{1}{8} \div 1\frac{3}{4} =$

23) $5\frac{1}{2} \div 2\frac{2}{5} =$

24) $3\frac{4}{7} \div 2\frac{2}{3} =$

25) $2\frac{4}{5} \div 3\frac{5}{6} =$

26) $2\frac{3}{7} \div 3\frac{2}{3} =$

Dividing Mixed Number – Answers

✎ *Solve and write the answer in lowest terms*

1) $9\frac{1}{2} \div 2\frac{3}{5} = 3\frac{17}{26}$

2) $2\frac{3}{8} \div 1\frac{2}{5} = 1\frac{39}{56}$

3) $5\frac{3}{4} \div 2\frac{2}{7} = 2\frac{33}{64}$

4) $8\frac{1}{3} \div 4\frac{1}{4} = 1\frac{49}{51}$

5) $7\frac{2}{5} \div 3\frac{3}{4} = 1\frac{73}{75}$

6) $2\frac{4}{5} \div 3\frac{2}{3} = \frac{42}{55}$

7) $8\frac{3}{5} \div 4\frac{3}{4} = 1\frac{77}{95}$

8) $6\frac{3}{4} \div 2\frac{2}{9} = 3\frac{3}{80}$

9) $5\frac{2}{7} \div 2\frac{2}{9} = 2\frac{53}{140}$

10) $2\frac{2}{5} \div 3\frac{3}{5} = \frac{2}{3}$

11) $4\frac{3}{7} \div 1\frac{7}{8} = 2\frac{38}{105}$

12) $2\frac{5}{7} \div 2\frac{4}{5} = \frac{95}{98}$

13) $8\frac{3}{5} \div 6\frac{1}{5} = 1\frac{12}{31}$

14) $2\frac{5}{8} \div 1\frac{8}{9} = 1\frac{53}{136}$

15) $5\frac{6}{7} \div 2\frac{3}{4} = 2\frac{10}{77}$

16) $1\frac{3}{5} \div 2\frac{3}{8} = \frac{64}{95}$

17) $5\frac{3}{4} \div 3\frac{2}{5} = 1\frac{47}{68}$

18) $2\frac{3}{4} \div 3\frac{1}{5} = \frac{55}{64}$

19) $3\frac{2}{3} \div 1\frac{2}{5} = 2\frac{13}{21}$

20) $4\frac{1}{4} \div 2\frac{2}{3} = 1\frac{19}{32}$

21) $3\frac{5}{6} \div 2\frac{4}{5} = 1\frac{31}{84}$

22) $2\frac{1}{8} \div 1\frac{3}{4} = 1\frac{3}{14}$

23) $5\frac{1}{2} \div 2\frac{2}{5} = 2\frac{7}{24}$

24) $3\frac{4}{7} \div 2\frac{2}{3} = 1\frac{19}{56}$

25) $2\frac{4}{5} \div 3\frac{5}{6} = \frac{84}{115}$

26) $2\frac{3}{7} \div 3\frac{2}{3} = \frac{51}{77}$

Comparing Decimals

✎ *Compare. Use >, =, and <*

1) 0.88 ☐ 0.088

2) 0.56 ☐ 0.57

3) 0.99 ☐ 0.89

4) 1.55 ☐ 1.65

5) 1.58 ☐ 1.75

6) 2.91 ☐ 2.85

7) 14.56 ☐ 1.456

8) 17.85 ☐ 17.89

9) 21.52 ☐ 21.052

10) 11.12 ☐ 11.03

11) 9.650 ☐ 9.65

12) 8.578 ☐ 8.568

13) 3.15 ☐ 0.315

14) 16.61 ☐ 16.16

15) 18.581 ☐ 8.991

16) 25.05 ☐ 2.505

17) 4.55 ☐ 4.65

18) 0.158 ☐ 1.58

19) 0.881 ☐ 0.871

20) 0.505 ☐ 0.510

21) 0.772 ☐ 0.777

22) 0.5 ☐ 0.500

23) 16.89 ☐ 15.89

24) 12.25 ☐ 12.35

25) 5.82 ☐ 5.69

26) 1.320 ☐ 1.032

27) 0.082 ☐ 0.088

28) 0.99 ☐ 0.099

29) 2.560 ☐ 1.950

30) 0.770 ☐ 0.707

31) 15.54 ☐ 1.554

32) 0.323 ☐ 0.332

Comparing Decimals – Answers

✍ *Compare. Use >, =, and <*

1) 0.88 > 0.088

2) 0.56 < 0.57

3) 0.99 > 0.89

4) 1.55 < 1.65

5) 1.58 < 1.75

6) 2.91 > 2.85

7) 14.56 > 1.456

8) 17.85 < 17.89

9) 21.52 > 21.052

10) 11.12 > 11.03

11) 9.650 = 9.65

12) 8.578 > 8.568

13) 3.15 > 0.315

14) 16.61 > 16.16

15) 18.581 > 8.991

16) 25.05 > 2.505

17) 4.55 < 4.65

18) 0.158 < 1.58

19) 0.881 > 0.871

20) 0.505 < 0.510

21) 0.772 < 0.777

22) 0.5 = 0.500

23) 16.89 > 15.89

24) 12.25 < 12.35

25) 5.82 > 5.69

26) 1.320 > 1.032

27) 0.082 < 0.088

28) 0.99 > 0.099

29) 2.560 > 1.950

30) 0.770 > 0.707

31) 15.54 > 1.554

32) 0.323 < 0.332

Rounding Decimals

✏️ *Round each number to the underlined place value.*

1) 2̲.814 =

2) 3.5̲62 =

3) 12.12̲5 =

4) 15̲.5 =

5) 1.9̲81 =

6) 14.2̲15 =

7) 17.54̲8 =

8) 25.50̲8 =

9) 31̲.089 =

10) 69.3̲45 =

11) 9.45̲7 =

12) 12̲.901 =

13) 2.65̲8 =

14) 32.5̲65 =

15) 6.05̲8 =

16) 98.10̲8 =

17) 27.7̲05 =

18) 36̲.75 =

19) 9.0̲8 =

20) 7.1̲85 =

21) 22.54̲7 =

22) 66.0̲98 =

23) 87̲.75 =

24) 18.5̲41 =

25) 10.25̲8 =

26) 13.4̲56 =

27) 71.08̲4 =

28) 29̲.23 =

29) 45.5̲5 =

30) 91̲.08 =

31) 83̲.433 =

32) 74.6̲4 =

bit.ly/3mKEluf

Find more at

Rounding Decimals - Answers

✍ *Round each number to the underlined place value.*

1) $\underline{2}.814 = 3$

2) $3.5\underline{6}2 = 3.56$

3) $12.1\underline{2}5 = 12.13$

4) $1\underline{5}.5 = 16$

5) $1.9\underline{8}1 = 1.98$

6) $14.\underline{2}15 = 14.2$

7) $17.5\underline{4}8 = 17.55$

8) $25.5\underline{0}8 = 25.51$

9) $3\underline{1}.089 = 31$

10) $69.\underline{3}45 = 69.3$

11) $9.4\underline{5}7 = 9.46$

12) $1\underline{2}.901 = 13$

13) $2.6\underline{5}8 = 2.66$

14) $32.\underline{5}65 = 32.6$

15) $6.0\underline{5}8 = 6.06$

16) $98.1\underline{0}8 = 98.11$

17) $27.\underline{7}05 = 27.7$

18) $3\underline{6}.75 = 37$

19) $9.\underline{0}8 = 9.1$

20) $7.\underline{1}85 = 7.2$

21) $22.5\underline{4}7 = 22.55$

22) $66.\underline{0}98 = 66.1$

23) $8\underline{7}.75 = 88$

24) $18.\underline{5}41 = 18.5$

25) $10.2\underline{5}8 = 10.26$

26) $13.\underline{4}56 = 13.5$

27) $71.0\underline{8}4 = 71.08$

28) $2\underline{9}.23 = 29$

29) $45.\underline{5}5 = 45.6$

30) $9\underline{1}.08 = 91$

31) $8\underline{3}.433 = 83$

32) $74.\underline{6}4 = 74.6$

Adding and Subtracting Decimals

✍ **Solve**.

1) $15.63 + 19.64 =$

2) $16.38 + 17.59 =$

3) $75.31 - 59.69 =$

4) $49.38 - 29.89 =$

5) $24.32 + 26.45 =$

6) $36.25 + 18.37 =$

7) $47.85 - 35.12 =$

8) $85.65 - 67.48 =$

9) $25.49 + 34.18 =$

10) $19.99 + 48.66 =$

11) $46.32 - 27.77 =$

12) $54.62 - 48.12 =$

13) $24.42 + 16.54 =$

14) $52.13 + 12.32 =$

15) $82.36 - 78.65 =$

16) $64.12 - 49.15 =$

17) $36.41 + 24.52 =$

18) $85.96 - 74.63 =$

19) $52.62 - 42.54 =$

20) $21.20 + 24.58 =$

21) $32.15 + 17.17 =$

22) $96.32 - 85.54 =$

23) $89.78 - 69.85 =$

24) $29.28 + 39.79 =$

25) $11.11 + 19.99 =$

26) $28.82 + 20.88 =$

27) $63.14 - 28.91 =$

28) $56.61 - 49.72 =$

29) $26.13 + 31.13 =$

30) $30.19 + 20.87 =$

31) $66.24 - 59.10 =$

32) $89.31 - 72.17 =$

bit.ly/38uyUpx   Find more at

Adding and Subtracting Decimals - Answers

✎ *Solve.*

1) $15.63 + 19.64 = 35.27$

2) $16.38 + 17.59 = 33.97$

3) $75.31 - 59.69 = 15.62$

4) $49.38 - 29.89 = 19.49$

5) $24.32 + 26.45 = 50.77$

6) $36.25 + 18.37 = 54.62$

7) $47.85 - 35.12 = 12.73$

8) $85.65 - 67.48 = 18.17$

9) $25.49 + 34.18 = 59.67$

10) $19.99 + 48.66 = 68.65$

11) $46.32 - 27.77 = 18.55$

12) $54.62 - 48.12 = 6.5$

13) $24.42 + 16.54 = 40.96$

14) $52.13 + 12.32 = 64.45$

15) $82.36 - 78.65 = 3.71$

16) $64.12 - 49.15 = 14.97$

17) $36.41 + 24.52 = 60.93$

18) $85.96 - 74.63 = 11.33$

19) $52.62 - 42.54 = 10.08$

20) $21.20 + 24.58 = 45.78$

21) $32.15 + 17.17 = 49.32$

22) $96.32 - 85.54 = 10.78$

23) $89.78 - 69.85 = 19.93$

24) $29.28 + 39.79 = 69.07$

25) $11.11 + 19.99 = 31.1$

26) $28.82 + 20.88 = 49.7$

27) $63.14 - 28.91 = 34.23$

28) $56.61 - 49.72 = 6.89$

29) $26.13 + 31.13 = 57.26$

30) $30.19 + 20.87 = 51.06$

31) $66.24 - 59.10 = 7.14$

32) $89.31 - 72.17 = 17.14$

Find more at

bit.ly/38uyUdx

Multiplying and Dividing Decimals

✎ *Solve.*

1) $11.2 \times 0.4 =$

2) $13.5 \times 0.8 =$

3) $42.2 \div 2 =$

4) $54.6 \div 6 =$

5) $23.1 \times 0.3 =$

6) $1.2 \times 0.7 =$

7) $5.5 \div 0.5 =$

8) $64.8 \div 8 =$

9) $1.4 \times 0.5 =$

10) $4.5 \times 0.3 =$

11) $88.8 \div 4 =$

12) $10.5 \div 5 =$

13) $2.2 \times 0.3 =$

14) $0.2 \times 0.52 =$

15) $95.7 \div 100 =$

16) $36.6 \div 6 =$

17) $3.2 \times 2 =$

18) $4.1 \times 0.5 =$

19) $68.4 \div 2 =$

20) $27.9 \div 9 =$

21) $3.5 \times 4 =$

22) $4.8 \times 0.5 =$

23) $6.4 \div 4 =$

24) $72.8 \div 0.8 =$

25) $1.8 \times 3 =$

26) $6.5 \times 0.2 =$

27) $93.6 \div 3 =$

28) $45.15 \div 0.5 =$

29) $13.2 \times 0.4 =$

30) $11.2 \times 5 =$

31) $7.2 \div 0.8 =$

32) $96.4 \div 0.2 =$

Multiplying and Dividing Decimals – Answers

✎ *Solve.*

1) $11.2 \times 0.4 = 4.48$

2) $13.5 \times 0.8 = 10.8$

3) $42.2 \div 2 = 21.1$

4) $54.6 \div 6 = 9.1$

5) $23.1 \times 0.3 = 6.93$

6) $1.2 \times 0.7 = 0.84$

7) $5.5 \div 5 = 1.1$

8) $64.8 \div 8 = 8.1$

9) $1.4 \times 0.5 = 0.7$

10) $4.5 \times 0.3 = 1.35$

11) $88.8 \div 4 = 22.2$

12) $10.5 \div 5 = 2.1$

13) $2.2 \times 0.3 = 0.66$

14) $0.2 \times 0.52 = 0.104$

15) $95.7 \div 100 = 0.957$

16) $36.6 \div 6 = 6.1$

17) $3.2 \times 2 = 6.4$

18) $4.1 \times 0.5 = 2.05$

19) $68.4 \div 2 = 34.2$

20) $27.9 \div 9 = 3.1$

21) $3.5 \times 4 = 14$

22) $4.8 \times 0.5 = 2.4$

23) $6.4 \div 4 = 1.6$

24) $72.8 \div 0.8 = 91$

25) $1.8 \times 3 = 5.4$

26) $6.5 \times 0.2 = 1.3$

27) $93.6 \div 3 = 31.2$

28) $45.15 \div 0.5 = 90.3$

29) $13.2 \times 0.4 = 5.28$

30) $11.2 \times 5 = 56$

31) $7.2 \div 0.8 = 9$

32) $96.4 \div 0.2 = 482$

Adding and Subtracting Integers

✍ *Solve.*

1) $-(8) + 13 =$

2) $17 - (-12 - 8) =$

3) $(-15) + (-4) =$

4) $(-14) + (-8) + 9 =$

5) $-(23) + 19 =$

6) $(-7 + 5) - 9 =$

7) $28 + (-32) =$

8) $(-11) + (-9) + 5 =$

9) $25 - (8 - 7) =$

10) $-(29) + 17 =$

11) $(-38) + (-3) + 29 =$

12) $15 - (-7 + 9) =$

13) $24 - (8 - 2) =$

14) $(-7 + 4) - 9 =$

15) $(-17) + (-3) + 9 =$

16) $(-26) + (-7) + 8 =$

17) $(-9) + (-11) =$

18) $8 - (-23 - 13) =$

19) $(-16) + (-2) =$

20) $25 - (7 - 4) =$

21) $23 + (-12) =$

22) $(-18) + (-6) =$

23) $17 - (-21 - 7) =$

24) $-(28) - (-16) + 5 =$

25) $(-9 + 4) - 8 =$

26) $(-28) + (-6) + 17 =$

27) $-(21) - (-15) + 9 =$

28) $(-31) + (-6) =$

29) $(-17) + (-11) + 14 =$

30) $(-29) + (-10) + 13 =$

31) $-(24) - (-12) + 5 =$

32) $8 - (-19 - 10) =$

Adding and Subtracting Integers – Answers

✍ *Solve*.

1) $-(8) + 13 = 5$

2) $17 - (-12 - 8) = 37$

3) $(-15) + (-4) = -19$

4) $(-14) + (-8) + 9 = -13$

5) $-(23) + 19 = -4$

6) $(-7 + 5) - 9 = -11$

7) $28 + (-32) = -4$

8) $(-11) + (-9) + 5 = -15$

9) $25 - (8 - 7) = 24$

10) $-(29) + 17 = -12$

11) $(-38) + (-3) + 29 = -12$

12) $15 - (-7 + 9) = 13$

13) $24 - (8 - 2) = 18$

14) $(-7 + 4) - 9 = -12$

15) $(-17) + (-3) + 9 = -11$

16) $(-26) + (-7) + 8 = -25$

17) $(-9) + (-11) = -20$

18) $8 - (-23 - 13) = 44$

19) $(-16) + (-2) = -18$

20) $25 - (7 - 4) = 22$

21) $23 + (-12) = 11$

22) $(-18) + (-6) = -24$

23) $17 - (-21 - 7) = 45$

24) $-(28) - (-16) + 5 = -7$

25) $(-9 + 4) - 8 = -13$

26) $(-28) + (-6) + 17 = -17$

27) $-(21) - (-15) + 9 = 3$

28) $(-31) + (-6) = -37$

29) $(-17) + (-11) + 14 = -14$

30) $(-29) + (-10) + 13 = -26$

31) $-(24) - (-12) + 5 = -7$

32) $8 - (-19 - 10) = 37$

Multiplying and Dividing Integers

✎ *Solve*.

1) $(-9) \times (-8) =$

2) $6 \times (-6) =$

3) $49 \div (-7) =$

4) $(-64) \div 8 =$

5) $(4) \times (-6) =$

6) $(-9) \times (-11) =$

7) $(10) \div (-5) =$

8) $144 \div (-12) =$

9) $(10) \times (-2) =$

10) $(-8) \times (-2) \times 5 =$

11) $(8) \div (-2) =$

12) $45 \div (-15) =$

13) $(5) \times (-7) =$

14) $(-6) \times (-5) \times 5 =$

15) $(12) \div (-6) =$

16) $(14) \div (-7) =$

17) $196 \div (-14) =$

18) $(27 - 13) \times (-2) =$

19) $125 \div (-5) =$

20) $66 \div (-6) =$

21) $(-6) \times (-5) \times 3 =$

22) $(15 - 6) \times (-3) =$

23) $(32 - 24) \div (-4) =$

24) $72 \div (-6) =$

25) $(-14 + 8) \times (-7) =$

26) $(-3) \times (-9) \times 3 =$

27) $84 \div (-12) =$

28) $(-12) \times (-10) =$

29) $25 \times (-4) =$

30) $(-3) \times (-5) \times 5 =$

31) $(15) \div (-3) =$

32) $(-18) \div (3) =$

bit.ly/3piQW98

Find more at

Multiplying and Dividing Integers - Answers

✎ *Solve.*

1) $(-9) \times (-8) = 72$

2) $6 \times (-6) = -36$

3) $49 \div (-7) = -7$

4) $(-64) \div 8 = -8$

5) $(4) \times (-6) = -24$

6) $(-9) \times (-11) = 99$

7) $(10) \div (-5) = -2$

8) $144 \div (-12) = -12$

9) $(10) \times (-2) = -20$

10) $(-8) \times (-2) \times 5 = 80$

11) $(8) \div (-2) = -4$

12) $45 \div (-15) = -3$

13) $(5) \times (-7) = -35$

14) $(-6) \times (-5) \times 5 = 150$

15) $(12) \div (-6) = -2$

16) $(14) \div (-7) = -2$

17) $196 \div (-14) = -14$

18) $(27 - 13) \times (-2) = -28$

19) $125 \div (-5) = -25$

20) $66 \div (-6) = -11$

21) $(-6) \times (-5) \times 3 = 90$

22) $(15 - 6) \times (-3) = -27$

23) $(32 - 24) \div (-4) = -2$

24) $72 \div (-6) = -12$

25) $(-14 + 8) \times (-7) = 42$

26) $(-3) \times (-9) \times 3 = 81$

27) $84 \div (-12) = -7$

28) $(-12) \times (-10) = 120$

29) $25 \times (-4) = -100$

30) $(-3) \times (-5) \times 5 = 75$

31) $(15) \div (-3) = -5$

32) $(-18) \div (3) = -6$

Order of Operation

✍ *Calculate.*

1) $18 + (32 \div 4) =$

2) $(3 \times 8) \div (-2) =$

3) $67 - (4 \times 8) =$

4) $(-11) \times (8 - 3) =$

5) $(18 - 7) \times (6) =$

6) $(6 \times 10) \div (12 + 3) =$

7) $(13 \times 2) - (24 \div 6) =$

8) $(-5) + (4 \times 3) + 8 =$

9) $(4 \times 2^3) + (16 - 9) =$

10) $(3^2 \times 7) \div (-2 + 1) =$

11) $[-2(48 \div 2^3)] - 6 =$

12) $(-4) + (7 \times 8) + 18 =$

13) $(3 \times 7) + (16 - 7) =$

14) $[3^3 \times (48 \div 2^3)] \div (-2) =$

15) $(14 \times 3) - (3^4 \div 9) =$

16) $(96 \div 12) \times (-3) =$

17) $(48 \div 2^2) \times (-2) =$

18) $(56 \div 7) \times (-5) =$

19) $(-2^2) + (7 \times 9) - 21 =$

20) $(2^4 - 9) \times (-6) =$

21) $[4^3 \times (50 \div 5^2)] \div (-16) =$

22) $(3^2 \times 4^2) \div (-4 + 2) =$

23) $6^2 - (-6 \times 4) + 3 =$

24) $4^2 - (5^2 \times 3) =$

25) $(-4) + (12^2 \div 3^2) - 7^2 =$

26) $(3^2 \times 5) + (-5^2 - 9) =$

27) $2[(3^2 \times 5) \times (-6)] =$

28) $(11^2 - 2^2) - (-7^2) =$

29) $(2^3 \times 3) - (49 \div 7) =$

30) $3[(3^2 \times 5) + (25 \div 5)] =$

31) $(6^2 \times 5) \div (-5) =$

32) $2^2[(6^3 \div 12) - (3^4 \div 27)] =$

Order of Operation – Answers

✎ *Calculate*.

1) $18 + (32 \div 4) = 26$

2) $(3 \times 8) \div (-2) = -12$

3) $67 - (4 \times 8) = 35$

4) $(-11) \times (8 - 3) = -55$

5) $(18 - 7) \times (6) = 66$

6) $(6 \times 10) \div (12 + 3) = 4$

7) $(13 \times 2) - (24 \div 6) = 22$

8) $(-5) + (4 \times 3) + 8 = 15$

9) $(4 \times 2^3) + (16 - 9) = 39$

10) $(3^2 \times 7) \div (-2 + 1) = -63$

11) $[-2(48 \div 2^3)] - 6 = -18$

12) $(-4) + (7 \times 8) + 18 = 70$

13) $(3 \times 7) + (16 - 7) = 30$

14) $[3^3 \times (48 \div 2^3)] \div (-2) = -81$

15) $(14 \times 3) - (3^4 \div 9) = 33$

16) $(96 \div 12) \times (-3) = -24$

17) $(48 \div 2^2) \times (-2) = -24$

18) $(56 \div 7) \times (-5) = -40$

19) $(-2^2) + (7 \times 9) - 21 = 38$

20) $(2^4 - 9) \times (-6) = -42$

21) $[4^3 \times (50 \div 5^2)] \div (-16) = -8$

22) $(3^2 \times 4^2) \div (-4 + 2) = -72$

23) $6^2 - (-6 \times 4) + 3 = 63$

24) $4^2 - (5^2 \times 3) = -59$

25) $(-4) + (12^2 \div 3^2) - 7^2 = -37$

26) $(3^2 \times 5) + (-5^2 - 9) = 11$

27) $2[(3^2 \times 5) \times (-6)] = -540$

28) $(11^2 - 2^2) - (-7^2) = 166$

29) $(2^3 \times 3) - (49 \div 7) = 17$

30) $3[(3^2 \times 5) + (25 \div 5)] = 150$

31) $(6^2 \times 5) \div (-5) = -36$

32) $2^2[(6^3 \div 12) - (3^4 \div 27)] = 60$

Integers and Absolute Value

✍ *Calculate*.

1) $5 - |8 - 12| =$

2) $|15| - \frac{|-16|}{4} =$

3) $\frac{|9 \times -6|}{18} \times \frac{|-24|}{8} =$

4) $|13 \times 3| + \frac{|-72|}{9} =$

5) $4 - |11 - 18| - |3| =$

6) $|18| - \frac{|-12|}{4} =$

7) $\frac{|5 \times -8|}{10} \times \frac{|-22|}{11} =$

8) $|9 \times 3| + \frac{|-36|}{4} =$

9) $|-42 + 7| \times \frac{|-2 \times 5|}{10} =$

10) $6 - |17 - 11| - |5| =$

11) $|13| - \frac{|-54|}{6} =$

12) $\frac{|9 \times -4|}{12} \times \frac{|-45|}{9} =$

13) $|-75 + 50| \times \frac{|-4 \times 5|}{5} =$

14) $\frac{|-26|}{13} \times \frac{|-32|}{8} =$

15) $14 - |8 - 18| - |-12| =$

16) $|29| - \frac{|-20|}{5} =$

17) $\frac{|3 \times 8|}{2} \times \frac{|-33|}{3} =$

18) $|-45 + 15| \times \frac{|-12 \times 5|}{6} =$

19) $\frac{|-50|}{5} \times \frac{|-77|}{11} =$

20) $12 - |2 - 7| - |15| =$

21) $|18| - \frac{|-45|}{15} =$

22) $\frac{|7 \times 8|}{4} \times \frac{|-48|}{12} =$

23) $\frac{|30 \times 2|}{3} \times |-12| =$

24) $\frac{|-36|}{9} \times \frac{|-80|}{8} =$

25) $|-35 + 8| \times \frac{|-9 \times 5|}{15} =$

26) $|19| - \frac{|-18|}{2} =$

27) $14 - |11 - 23| + |2| =$

28) $|-39 + 7| \times \frac{|-4 \times 6|}{3} =$

Integers and Absolute Value - Answers

✍ *Calculate*.

1) $5 - |8 - 12| = 1$

2) $|15| - \frac{|-16|}{4} = 11$

3) $\frac{|9 \times -6|}{18} \times \frac{|-24|}{8} = 9$

4) $|13 \times 3| + \frac{|-72|}{9} = 47$

5) $4 - |11 - 18| - |3| = -6$

6) $|18| - \frac{|-12|}{4} = 15$

7) $\frac{|5 \times -8|}{10} \times \frac{|-22|}{11} = 8$

8) $|9 \times 3| + \frac{|-36|}{4} = 36$

9) $|-42 + 7| \times \frac{|-2 \times 5|}{10} = 35$

10) $6 - |17 - 11| - |5| = -5$

11) $|13| - \frac{|-54|}{6} = 4$

12) $\frac{|9 \times -4|}{12} \times \frac{|-45|}{9} = 15$

13) $|-75 + 50| \times \frac{|-4 \times 5|}{5} = 100$

14) $\frac{|-26|}{13} \times \frac{|-32|}{8} = 8$

15) $14 - |8 - 18| - |-12| = -8$

16) $|29| - \frac{|-20|}{5} = 25$

17) $\frac{|3 \times 8|}{2} \times \frac{|-33|}{3} = 132$

18) $|-45 + 15| \times \frac{|-12 \times 5|}{6} = 300$

19) $\frac{|-50|}{5} \times \frac{|-77|}{11} = 70$

20) $12 - |2 - 7| - |15| = -8$

21) $|18| - \frac{|-45|}{15} = 15$

22) $\frac{|7 \times 8|}{4} \times \frac{|-48|}{12} = 56$

23) $\frac{|30 \times 2|}{3} \times |-12| = 240$

24) $\frac{|-36|}{9} \times \frac{|-80|}{8} = 40$

25) $|-35 + 8| \times \frac{|-9 \times 5|}{15} = 81$

26) $|19| - \frac{|-18|}{2} = 10$

27) $14 - |11 - 23| + |2| = 4$

28) $|-39 + 7| \times \frac{|-4 \times 6|}{3} = 256$

Simplifying Ratios

✍ *Simplify each ratio*.

1) $3 : 27 = $ ___ : ___

2) $2 : 8 = $ ___ : ___

3) $\frac{4}{28} = -$

4) $\frac{16}{40} = -$

5) $10 : 30 = $ ___ : ___

6) $5 : 30 = $ ___ : ___

7) $\frac{34}{38} = -$

8) $\frac{45}{63} = -$

9) $10 : 45 = $ ___ : ___

10) $20 : 30 = $ ___ : ___

11) $\frac{40}{64} = -$

12) $\frac{10}{110} = -$

13) $8 : 12 = $ ___ : ___

14) $16 : 20 = $ ___ : ___

15) $\frac{24}{48} = -$

16) $\frac{21}{77} = -$

17) $8 : 24 = $ ___ : ___

18) 9 to $36 = $ ___ : ___

19) $\frac{64}{72} = -$

20) $\frac{45}{60} = -$

21) $12 : 15 = $ ___ : ___

22) $18 : 54 = $ ___ · ___

23) $\frac{36}{54} = -$

24) $\frac{48}{104} = -$

25) $15 : 75 = $ ___ : ___

26) $16 : 48 = $ ___ : ___

27) $\frac{15}{65} = -$

28) $\frac{44}{52} = -$

Simplifying Ratios – Answers

✎ *Simplify each ratio.*

1) $3 : 27 = 1 : 9$

2) $2 : 8 = 1 : 4$

3) $\frac{4}{28} = \frac{1}{7}$

4) $\frac{16}{40} = \frac{2}{5}$

5) $10 : 30 = 1 : 3$

6) $5 : 30 = 1 : 6$

7) $\frac{34}{38} = \frac{17}{19}$

8) $\frac{45}{63} = \frac{5}{7}$

9) $10 : 45 = 2 : 9$

10) $20 : 30 = 2 : 3$

11) $\frac{40}{64} = \frac{5}{8}$

12) $\frac{10}{110} = \frac{1}{11}$

13) $8 : 12 = 2 : 3$

14) $16 : 20 = 4 : 5$

15) $\frac{24}{48} = \frac{1}{2}$

16) $\frac{21}{77} = \frac{3}{11}$

17) $8 : 24 = 1 : 3$

18) $9 \text{ to } 36 = 1 \text{ to } 4$

19) $\frac{64}{72} = \frac{8}{9}$

20) $\frac{45}{60} = \frac{3}{4}$

21) $12 : 15 = 4 : 5$

22) $18 : 54 = 1 : 3$

23) $\frac{36}{54} = \frac{2}{3}$

24) $\frac{48}{104} = \frac{6}{13}$

25) $15 : 75 = 1 : 5$

26) $16 : 48 = 1 : 3$

27) $\frac{15}{65} = \frac{3}{13}$

28) $\frac{44}{52} = \frac{11}{13}$

bit.ly/3nKwg0Z

Proportional Ratios

 *Solve each proportion for x.*

1) $\frac{4}{7} = \frac{16}{x}$, $x =$ _____

2) $\frac{4}{9} = \frac{x}{18}$, $x =$ _____

3) $\frac{3}{5} = \frac{24}{x}$, $x =$ _____

4) $\frac{3}{10} = \frac{x}{50}$, $x =$ _____

5) $\frac{3}{11} = \frac{15}{x}$, $x =$ _____

6) $\frac{6}{15} = \frac{x}{45}$, $x =$ _____

7) $\frac{6}{19} = \frac{12}{x}$, $x =$ _____

8) $\frac{7}{16} = \frac{x}{32}$, $x =$ _____

9) $\frac{18}{21} = \frac{54}{x}$, $x =$ _____

10) $\frac{13}{15} = \frac{39}{x}$, $x =$ _____

11) $\frac{9}{13} = \frac{72}{x}$, $x =$ _____

12) $\frac{8}{30} = \frac{x}{180}$, $x =$ _____

13) $\frac{3}{19} = \frac{9}{x}$, $x =$ _____

14) $\frac{1}{3} = \frac{x}{90}$, $x =$ _____

15) $\frac{25}{45} = \frac{x}{9}$, $x =$ _____

16) $\frac{1}{6} = \frac{9}{x}$, $x =$ _____

17) $\frac{7}{9} = \frac{63}{x}$, $x =$ _____

18) $\frac{54}{72} = \frac{x}{8}$, $x =$ _____

19) $\frac{32}{40} = \frac{4}{x}$, $x =$ _____

20) $\frac{21}{42} = \frac{x}{6}$, $x =$ _____

21) $\frac{56}{72} = \frac{7}{x}$, $x =$ _____

22) $\frac{1}{14} = \frac{x}{42}$, $x =$ _____

23) $\frac{5}{7} = \frac{75}{x}$, $x =$ _____

24) $\frac{30}{48} = \frac{x}{8}$, $x =$ _____

25) $\frac{36}{88} = \frac{9}{x}$, $x =$ _____

26) $\frac{62}{68} = \frac{x}{34}$, $x =$ _____

27) $\frac{42}{60} = \frac{x}{10}$, $x =$ _____

28) $\frac{8}{9} = \frac{x}{108}$, $x =$ _____

29) $\frac{46}{69} = \frac{x}{3}$, $x =$ _____

30) $\frac{99}{121} = \frac{x}{11}$, $x =$ _____

31) $\frac{19}{21} = \frac{x}{63}$, $x =$ _____

32) $\frac{11}{12} = \frac{x}{48}$, $x =$ _____

bit.ly/37GHOxp

Find more

Proportional Ratios - Answers

✍ *Solve each proportion for x.*

1) $\frac{4}{7} = \frac{16}{x}, x = 28$

2) $\frac{4}{9} = \frac{x}{18}, x = 8$

3) $\frac{3}{5} = \frac{24}{x}, x = 40$

4) $\frac{3}{10} = \frac{x}{50}, x = 15$

5) $\frac{3}{11} = \frac{15}{x}, x = 55$

6) $\frac{6}{15} = \frac{x}{45}, x = 18$

7) $\frac{6}{19} = \frac{12}{x}, x = 38$

8) $\frac{7}{16} = \frac{x}{32}, x = 14$

9) $\frac{18}{21} = \frac{54}{x}, x = 63$

10) $\frac{13}{15} = \frac{39}{x}, x = 45$

11) $\frac{9}{13} = \frac{72}{x}, x = 104$

12) $\frac{8}{30} = \frac{x}{180}, x = 48$

13) $\frac{3}{19} = \frac{9}{x}, x = 57$

14) $\frac{1}{3} = \frac{x}{90}, x = 30$

15) $\frac{25}{45} = \frac{x}{9}, x = 5$

16) $\frac{1}{6} = \frac{9}{x}, x = 54$

17) $\frac{7}{9} = \frac{63}{x}, x = 81$

18) $\frac{54}{72} = \frac{x}{8}, x = 6$

19) $\frac{32}{40} = \frac{4}{x}, x = 5$

20) $\frac{21}{42} = \frac{x}{6}, x = 3$

21) $\frac{56}{72} = \frac{7}{x}, x = 9$

22) $\frac{1}{14} = \frac{x}{42}, x = 3$

23) $\frac{5}{7} = \frac{75}{x}, x = 105$

24) $\frac{30}{48} = \frac{x}{8}, x = 5$

25) $\frac{36}{88} = \frac{9}{x}, x = 22$

26) $\frac{62}{68} = \frac{x}{34}, x = 31$

27) $\frac{42}{60} = \frac{x}{10}, x = 7$

28) $\frac{8}{9} = \frac{x}{108}, x = 96$

29) $\frac{46}{69} = \frac{x}{3}, x = 2$

30) $\frac{99}{121} = \frac{x}{11}, x = 9$

31) $\frac{19}{21} = \frac{x}{63}, x = 57$

32) $\frac{11}{12} = \frac{x}{48}, x = 44$

Create Proportion

✎ *State if each pair of ratios form a proportion.*

1) $\frac{5}{8}$ and $\frac{25}{50}$

2) $\frac{2}{11}$ and $\frac{4}{22}$

3) $\frac{2}{5}$ and $\frac{8}{20}$

4) $\frac{3}{11}$ and $\frac{9}{33}$

5) $\frac{5}{10}$ and $\frac{15}{30}$

6) $\frac{4}{13}$ and $\frac{8}{24}$

7) $\frac{6}{9}$ and $\frac{24}{36}$

8) $\frac{7}{12}$ and $\frac{14}{20}$

9) $\frac{3}{8}$ and $\frac{27}{72}$

10) $\frac{12}{20}$ and $\frac{36}{60}$

11) $\frac{11}{12}$ and $\frac{55}{60}$

12) $\frac{12}{15}$ and $\frac{24}{25}$

13) $\frac{15}{19}$ and $\frac{20}{38}$

14) $\frac{10}{14}$ and $\frac{40}{56}$

15) $\frac{11}{13}$ and $\frac{44}{39}$

16) $\frac{15}{16}$ and $\frac{30}{32}$

17) $\frac{17}{19}$ and $\frac{34}{48}$

18) $\frac{5}{18}$ and $\frac{15}{54}$

19) $\frac{3}{14}$ and $\frac{18}{42}$

20) $\frac{7}{11}$ and $\frac{14}{32}$

21) $\frac{8}{11}$ and $\frac{32}{44}$

22) $\frac{9}{13}$ and $\frac{18}{26}$

✎ *Solve.*

23) The ratio of boys to girls in a class is 5:6. If there are 25 boys in the class, how many girls are in that class? _____

24) The ratio of red marbles to blue marbles in a bag is 4:7. If there are 77 marbles in the bag, how many of the marbles are red? _____

25) You can buy 8 cans of green beans at a supermarket for $3.20. How much does it cost to buy 48 cans of green beans? _____

Create Proportion – Answers

State if each pair of ratios form a proportion.

1) $\frac{5}{8}$ and $\frac{25}{50}$, *No*

2) $\frac{2}{11}$ and $\frac{4}{22}$, *Yes*

3) $\frac{2}{5}$ and $\frac{8}{20}$, *Yes*

4) $\frac{3}{11}$ and $\frac{9}{33}$, *Yes*

5) $\frac{5}{10}$ and $\frac{15}{30}$, *Yes*

6) $\frac{4}{13}$ and $\frac{8}{24}$, *No*

7) $\frac{6}{9}$ and $\frac{24}{36}$, *Yes*

8) $\frac{7}{12}$ and $\frac{14}{20}$, *No*

9) $\frac{3}{8}$ and $\frac{27}{72}$, *Yes*

10) $\frac{12}{20}$ and $\frac{36}{60}$, *Yes*

11) $\frac{11}{12}$ and $\frac{55}{60}$, *Yes*

12) $\frac{12}{15}$ and $\frac{24}{25}$, *No*

13) $\frac{15}{19}$ and $\frac{20}{38}$, *No*

14) $\frac{10}{14}$ and $\frac{40}{56}$, *Yes*

15) $\frac{11}{13}$ and $\frac{44}{39}$, *No*

16) $\frac{15}{16}$ and $\frac{30}{32}$, *Yes*

17) $\frac{17}{19}$ and $\frac{34}{38}$, *Yes*

18) $\frac{5}{18}$ and $\frac{15}{54}$, *Yes*

19) $\frac{3}{14}$ and $\frac{18}{42}$, *No*

20) $\frac{7}{11}$ and $\frac{14}{32}$, *No*

21) $\frac{8}{11}$ and $\frac{32}{44}$, *Yes*

22) $\frac{9}{13}$ and $\frac{18}{26}$, *Yes*

Solve.

23) The ratio of boys to girls in a class is $5:6$. If there are 25 boys in the class, how many girls are in that class? **30 girls**

24) The ratio of red marbles to blue marbles in a bag is $4:7$. If there are 77 marbles in the bag, how many of the marbles are red? **28 red marbles**

25) You can buy 8 cans of green beans at a supermarket for $3.20. How much does it cost to buy 48 cans of green beans? **$19.20**

Similarity and Ratios

✎ *Each pair of figures is similar. Find the missing side.*

1)

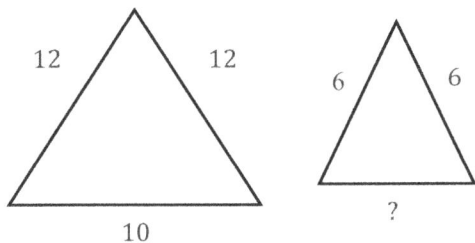

2)

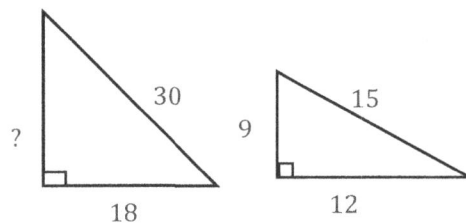

3)

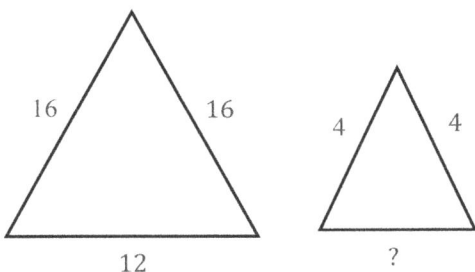

4)

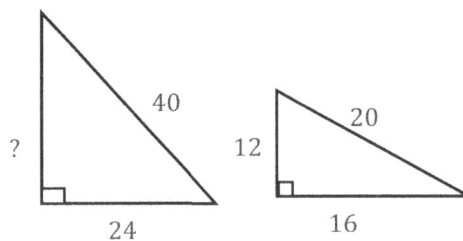

5)

6)

7)

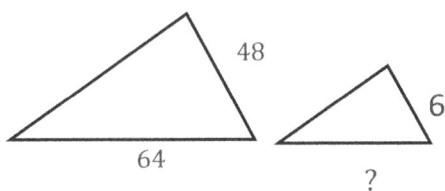

8)

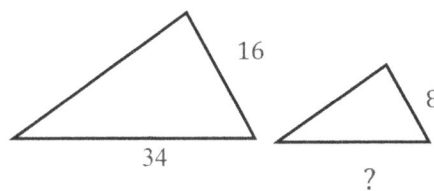

Similarity and Ratios – Answers

Each pair of figures is similar. Find the missing side.

1) **5**

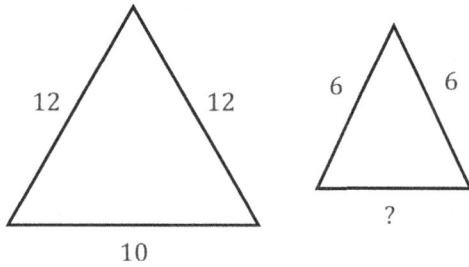

2) **24**

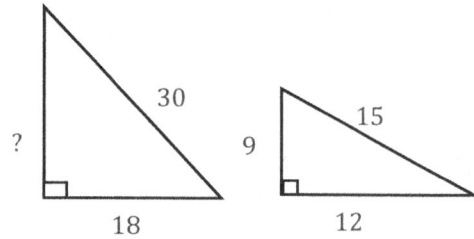

3) **3**

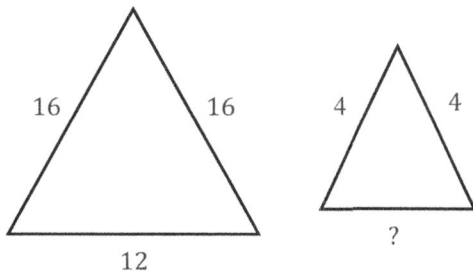

4) **32**

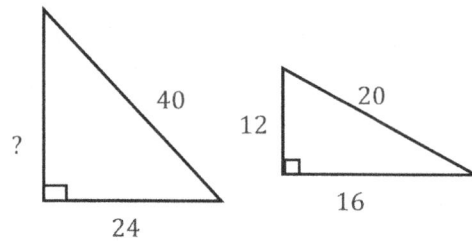

5) **9**

6) **8**

7) **8**

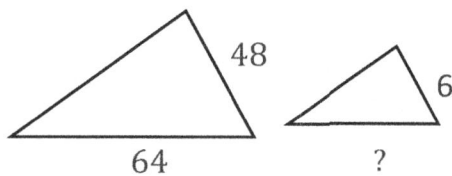

8) **17**

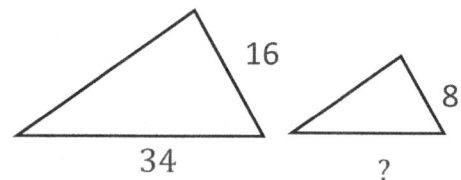

EffortlessMath.com

Simple Interest

✍ *Determine the simple interest for the following loans.*

1) $440 at 5% for 6 years. $___
2) $460 at 2.5% for 4 years. $_
3) $500 at 3% for 5 years. $___
4) $550 at 9% for 2 years. $___
5) $690 at 5% for 6 months. $___
6) $620 at 7% for 3 years. $___
7) $650 at 4.5% for 10 years. $___
8) $850 at 4% for 2 years. $___
9) $640 at 7% for 3 years. $___
10) $300 at 9% for 9 months. $___
11) $760 at 8% for 2 years. $_
12) $910 at 5% for 5 years. $___
13) $540 at 3% for 6 years. $___
14) $780 at 2.5% for 4 years. $___
15) $1,600 at 7% for 3 months. $___
16) $310 at 4% for 4 years. $___
17) $950 at 6% for 5 years. $___
18) $280 at 8% for 7 years. $___
19) $310 at 6% for 3 years. $___
20) $990 at 5% for 4 months. $___
21) $380 at 6% for 5 years. $___

22) $580 at 6% for 4 years. $___
23) $1,200 at 4% for 5 years. $___
24) $3,100 at 5% for 6 years. $___
25) $5,200 at 8% for 2 years. $___
26) $1,400 at 4% for 3 years. $___
27) $300 at 3% for 8 months. $___
28) $150 at 3.5% for 4 years. $___
29) $170 at 6% for 2 years. $___
30) $940 at 8% for 5 years. $___
31) $960 at 1.5% for 8 years. $_
32) $240 at 5% for 4 months. $___
33) $280 at 2% for 5 years. $___
34) $880 at 3% for 2 years. $___
35) $2,200 at 4.5% for 2 years. $___
36) $2,400 at 7% for 3 years. $___
37) $1,800 at 5% for 6 months. $___
38) $190 at 4% for 2 years. $___
39) $560 at 7% for 4 years. $___
40) $720 at 8% for 2 years. $_
41) $780 at 5% for 8 years. $___
42) $880 at 6% for 3 months. $___

Simple Interest - Answers

Determine the simple interest for the following loans.

1) $440 at 5% for 6 years. $132
2) $460 at 2.5% for 4 years. $46
3) $500 at 3% for 5 years. $75
4) $550 at 9% for 2 years. $99
5) $690 at 5% for 6 months. $17.25
6) $620 at 7% for 3 years. $130.20
7) $650 at 4.5% for 10 years. $292.50
8) $850 at 4% for 2 years. $68
9) $640 at 7% for 3 years. $134.40
10) $300 at 9% for 9 months. $20.25
11) $760 at 8% for 2 years. $121.60
12) $910 at 5% for 5 years. $227.50
13) $540 at 3% for 6 years. $97.20
14) $780 at 2.5% for 4 years. $78
15) $1,600 at 7% for 3 months. $28
16) $310 at 4% for 4 years. $49.60
17) $950 at 6% for 5 years. $285
18) $280 at 8% for 7 years. $156.80
19) $310 at 6% for 3 years. $55.80
20) $990 at 5% for 4 months. $198
21) $380 at 6% for 5 years. $114

22) $580 at 6% for 4 years. $139.20
23) $1,200 at 4% for 5 years. $240
24) $3,100 at 5% for 6 years. $930
25) $5,200 at 8% for 2 years. $832
26) $1,400 at 4% for 3 years. $168
27) $300 at 3% for 8 months. $6
28) $150 at 3.5% for 4 years. $21
29) $170 at 6% for 2 years. $16.5
30) $940 at 8% for 5 years. $376
31) $960 at 1.5% for 8 years. $115.20
32) $240 at 5% for 4 months. $4
33) $280 at 2% for 5 years. $28
34) $880 at 3% for 2 years. $52.80
35) $2,200 at 4.5% for 2 years. $198
36) $2,400 at 7% for 3 years. $504
37) $1,800 at 5% for 6 months. $45
38) $190 at 4% for 2 years. $15.20
39) $560 at 7% for 4 years. $156.80
40) $720 at 8% for 2 years. $115.20
41) $780 at 5% for 8 years. $312
42) $880 at 6% for 3 months. $13.20

Percent Problems

✍ *Solve each problem.*

1) What is 5 percent of 300? ____

2) What is 15 percent of 600? ____

3) What is 12 percent of 450? ____

4) What is 30 percent of 240? ____

5) What is 60 percent of 850? ____

6) 63 is what percent of 300? ____%

7) 80 is what percent of 400? ____%

8) 70 is what percent of 700? ____%

9) 84 is what percent of 600? ____%

10) 90 is what percent of 300? ____%

11) 24 is what percent of 150? ____%

12) 12 is what percent of 80? ____%

13) 4 is what percent of 50? ____%

14) 110 is what percent of 500? __%

15) 16 is what percent of 400? __%

16) 39 is what percent of 300? ____%

17) 56 is what percent of 200? ____%

18) 30 is what percent of 500? ____%

19) 84 is what percent of 700? ____%

20) 40 is what percent of 500? __%

21) 26 is what percent of 100? __ %

22) 45 is what percent of 900? __%

23) 60 is what percent of 400? ____%

24) 18 is what percent of 900? ____%

25) 75 is what percent of 250? ____%

26) 27 is what percent of 900? ____%

27) 49 is what percent of 700? ____%

28) 81 is what percent of 900? ____%

29) 90 is what percent of 500? ____%

30) 82 is 20 percent of what number? ____

31) 14 is 35 percent of what number? ____

32) 90 is 6 percent of what number? ____

33) 80 is 40 percent of what number? ____

34) 90 is 15 percent of what number? ____

35) 28 is 7 percent of what number? ____

36) 54 is 18 percent of what number? ____

37) 72 is 24 percent of what number? ____

Percent Problems - Answers

 **Solve each problem.**

1) What is 5 percent of 300? 15

2) What is 15 percent of 600? 90

3) What is 12 percent of 450? 54

4) What is 30 percent of 240? 72

5) What is 60 percent of 850? 510

6) 63 is what percent of 300? 21%

7) 80 is what percent of 400? 20%

8) 70 is what percent of 700? 10%

9) 84 is what percent of 600? 14%

10) 90 is what percent of 300? 30%

11) 24 is what percent of 150? 16%

12) 12 is what percent of 80? 15%

13) 4 is what percent of 50? 8%

14) 110 is what percent of 500? 22%

15) 16 is what percent of 400? 4%

16) 39 is what percent of 300? 13%

17) 56 is what percent of 200? 28%

18) 30 is what percent of 500? 6%

19) 84 is what percent of 700? 12%

20) 40 is what percent of 500? 8%

21) 26 is what percent of 100? 26%

22) 45 is what percent of 900? 5%

23) 60 is what percent of 400? 15%

24) 18 is what percent of 900? 2%

25) 75 is what percent of 250? 30%

26) 27 is what percent of 900? 3%

27) 49 is what percent of 700? 7%

28) 81 is what percent of 900? 9%

29) 90 is what percent of 500? 18%

30) 82 is 20 percent of what number? 410

31) 14 is 35 percent of what number? 40

32) 90 is 6 percent of what number? 1,500

33) 80 is 40 percent of what number? 200

34) 90 is 15 percent of what number? 600

35) 28 is 7 percent of what number? 400

36) 54 is 18 percent of what number? 300

37) 72 is 24 percent of what number? 300

Percent of Increase and Decrease

✍ *Solve each percent of the change word problem*.

1) Bob got a raise, and his hourly wage increased from $24 to $36. What is the percent increase? _____

2) The price of gasoline rose from $2.20 to $2.42 in one month. By what percent did the gas price rise? _____

3) In a class, the number of students has been increased from 30 to 39. What is the percent increase? _____

4) The price of a pair of shoes increases from $28 to $35. What is the percent increase? _____

5) In a class, the number of students has been decreased from 24 to 18. What is the percentage decrease? _____

6) Nick got a raise, and his hourly wage increased from $50 to $55. What is the percent increase? _____

7) A coat was originally priced at $80. It went on sale for $70.40. What was the percent that the coat was discounted? ____

8) The price of a pair of shoes increases from $8 to $12. What is the percent increase? _____

9) A house was purchased in 2002 for $180,000. It is now valued at $144,000. What is the rate (percent) of depreciation for the house? _____

10) The price of gasoline rose from $3.00 to $3.15 in one month. By what percent did the gas price rise? _____

Percent of Increase and Decrease – Answers

Solve each percent of the change word problem

1) Bob got a raise, and his hourly wage increased from $24 to $36. What is the percent increase? 50%

2) The price of gasoline rose from $2.20 to $2.42 in one month. By what percent did the gas price rise? 10%

3) In a class, the number of students has been increased from 30 to 39. What is the percent increase? 30%

4) The price of a pair of shoes increases from $28 to $35. What is the percent increase? 25%

5) In a class, the number of students has been decreased from 24 to 18. What is the percentage decrease? 25%

6) Nick got a raise, and his hourly wage increased from $50 to $55. What is the percent increase? 10%

7) A coat was originally priced at $80. It went on sale for $70.40. What was the percent that the coat was discounted? 12%

8) The price of a pair of shoes increases from $8 to $12. What is the percent increase? 50%

9) A house was purchased in 2002 for $180,000. It is now valued at $144,000. What is the rate (percent) of depreciation for the house? 20%

10) The price of gasoline rose from $3.00 to $3.15 in one month. By what percent did the gas price rise? 5%

Discount, Tax and Tip

✎ *Find the missing values.*

1) Original price of a computer: $400, Tax: 5%, Selling price: $_____

2) Original price of a sofa: $600, Tax: 12%, Selling price: $_____

3) Original price of a table: $550, Tax: 18%, Selling price: $_____

4) Original price of a cell phone: $700, Tax: 20%, Selling price: $_____

5) Original price of a printer: $400, Tax: 22%, Selling price: $_____

6) Original price of a computer: $600, Tax: 15%, Selling price: $_____

7) Restaurant bill: $24.00, Tip: 25%, Final amount: $_____

8) Original price of a cell phone: $300 Tax: 8%, Selling price: $_____

9) Original price of a carpet: $800, Tax: 25%, Selling price: $_____

10) Original price of a camera: $200 Discount: 35%, Selling price: $_____

11) Original price of a dress: $500 Discount: 10%, Selling price: $_____

12) Original price of a monitor: $400 Discount: 5%, Selling price: $_____

13) Original price of a laptop: $900 Discount: 20%, Selling price: $_____

14) Restaurant bill: $54.00 Tip: 20%, Final amount: $_____

EffortlessMath.com

bit.ly/2Je5lo0

Find more at

Discount, Tax and Tip – Answers

✎ *Find the missing values.*

1) Original price of a computer: $400 Tax: 5%, Selling price: $420

2) Original price of a sofa: $600 Tax: 12%, Selling price: $672

3) Original price of a table: $550 Tax: 18%, Selling price: $649

4) Original price of a cell phone: $700 Tax: 20%, Selling price: $840

5) Original price of a printer: $400 Tax: 22%, Selling price: $488

6) Original price of a computer: $600 Tax: 15%, Selling price: $690

7) Restaurant bill: $24.00 Tip: 25%, Final amount: $30.00

8) Original price of a cell phone: $300 Tax: 8%, Selling price: $324

9) Original price of a carpet: $800 Tax: 25%, Selling price: $1,000

10) Original price of a camera: $200 Discount: 35%, Selling price: $130

11) Original price of a dress: $500 Discount: 10%, Selling price: $450

12) Original price of a monitor: $400 Discount: 5%, Selling price: $380

13) Original price of a laptop: $900 Discount: 20%, Selling price: $720

14) Restaurant bill: $54.00 Tip: 20%, Final amount: $64.80

Simplifying Variable Expressions

✍️ *Simplify and write the answer.*

1) $3x + 5 + 2x =$

2) $7x + 3 - 3x =$

3) $-2 - x^2 - 6x^2 =$

4) $(-6)(8x - 4) =$

5) $3 + 10x^2 + 2x =$

6) $8x^2 + 6x + 7x^2 =$

7) $2x^2 - 5x - 7x =$

8) $x - 3 + 5 - 3x =$

9) $2 - 3x + 12 - 2x =$

10) $5x^2 - 12x^2 + 8x =$

11) $2x^2 + 6x + 3x^2 =$

12) $2x^2 - 2x - x =$

13) $2x^2 - (-8x + 6) = 2$

14) $4x + 6(2 - 5x) =$

15) $10x + 8(10x - 6) =$

16) $9(-2x - 6) - 5 =$

17) $32x - 4 + 23 + 2x =$

18) $8x - 12x - x^2 + 13 =$

19) $(-6)(8x - 4) + 10x =$

20) $14x - 5(5 - 8x) =$

21) $23x + 4(9x + 3) + 12 =$

22) $3(-7x + 5) + 20x =$

23) $12x - 3x(x + 9) =$

24) $7x + 5x(3 - 3x) =$

25) $5x(-8x + 12) + 14x =$

26) $40x + 12 + 2x^2 =$

27) $5x(x - 3) - 10 =$

28) $8x - 7 + 8x + 2x^2 =$

29) $7x - 3x^2 - 5x^2 - 3 =$

30) $4 + x^2 - 6x^2 - 12x =$

31) $12x + 8x^2 + 2x + 20 =$

32) $23 + 15x^2 + 8x - 4x^2 =$

Simplifying Variable Expressions - Answers

✎ *Simplify and write the answer.*

1) $3x + 5 + 2x = 5x + 5$

2) $7x + 3 - 3x = 4x + 3$

3) $-2 - x^2 - 6x^2 = -7x^2 - 2$

4) $(-6)(8x - 4) = -48x + 24$

5) $3 + 10x^2 + 2x = 10x^2 + 2x + 3$

6) $8x^2 + 6x + 7x^2 = 15x^2 + 6x$

7) $2x^2 - 5x - 7x = 2x^2 - 12x$

8) $x - 3 + 5 - 3x = -2x + 2$

9) $2 - 3x + 12 - 2x = -5x + 14$

10) $5x^2 - 12x^2 + 8x = -7x^2 + 8x$

11) $2x^2 + 6x + 3x^2 = 5x^2 + 6x$

12) $2x^2 - 2x - x = 2x^2 - 3x$

13) $2x^2 - (-8x + 6) = 2x^2 + 8x - 6$

14) $4x + 6(2 - 5x) = -26x + 12$

15) $10x + 8(10x - 6) = 90x - 48$

16) $9(-2x - 6) - 5 = -18x - 59$

17) $32x - 4 + 23 + 2x = 34x + 19$

18) $8x - 12x - x^2 + 13 = -x^2 - 4x + 13$

19) $(-6)(8x - 4) + 10x = -38x + 24$

20) $14x - 5(5 - 8x) = 54x - 25$

21) $23x + 4(9x + 3) + 12 = 59x + 24$

22) $3(-7x + 5) + 20x = -x + 15$

23) $12x - 3x(x + 9) = -3x^2 - 15x$

24) $7x + 5x(3 - 3x) = -15x^2 + 22x$

25) $5x(-8x + 12) + 14x = -40x^2 + 74x$

26) $40x + 12 + 2x^2 = 2x^2 + 40x + 12$

27) $5x(x - 3) - 10 = 5x^2 - 15x - 10$

28) $8x - 7 + 8x + 2x^2 = 2x^2 + 16x - 7$

29) $7x - 3x^2 - 5x^2 - 3 = -8x^2 + 7x - 3$

30) $4 + x^2 - 6x^2 - 12x = -5x^2 - 12x + 4$

31) $12x + 8x^2 + 2x + 20 = 8x^2 + 14x + 20$

32) $23 + 15x^2 + 8x - 4x^2 = 11x^2 + 8x + 23$

Simplifying Polynomial Expressions

✎ *Simplify and write the answer.*

1) $(2x^3 + 5x^2) - (12x + 2x^2) =$ _____

2) $(-x^5 + 2x^3) - (3x^3 + 6x^2) =$ _____

3) $(12x^4 + 4x^2) - (2x^2 - 6x^4) =$ _____

4) $4x - 3x^2 - 2(6x^2 + 6x^3) =$ _____

5) $(2x^3 - 3) + 3(2x^2 - 3x^3) =$ _____

6) $4(4x^3 - 2x) - (3x^3 - 2x^4) =$ _____

7) $2(4x - 3x^3) - 3(3x^3 + 4x^2) =$ _____

8) $(2x^2 - 2x) - (2x^3 + 5x^2) =$ _____

9) $2x^3 - (4x^4 + 2x) + x^2 =$ _____

10) $x^4 - 9(x^2 + x) - 5x =$ _____

11) $(-2x^2 - x^4) + (4x^4 - x^2) =$ _____

12) $4x^2 - 5x^3 + 15x^4 - 12x^3 =$ _____

13) $2x^2 - 5x^4 + 14x^4 - 11x^3 =$ _____

14) $2x^2 + 5x^3 - 7x^2 + 12x =$ _____

15) $2x^4 - 5x^5 + 8x^4 - 8x^2 =$ _____

16) $5x^3 + 17x - 5x^2 - 2x^3 =$ _____

Simplifying Polynomial Expressions - Answers

✎ *Simplify and write the answer.*

1) $(2x^3 + 5x^2) - (12x + 2x^2) = 2x^3 + 3x^2 - 12x$

2) $(-x^5 + 2x^3) - (3x^3 + 6x^2) = -x^5 - x^3 - 6x^2$

3) $(12x^4 + 4x^2) - (2x^2 - 6x^4) = 18x^4 + 2x^2$

4) $4x - 3x^2 - 2(6x^2 + 6x^3) = -12x^3 - 15x^2 + 4x$

5) $(2x^3 - 3) + 3(2x^2 - 3x^3) = -7x^3 + 6x^2 - 3$

6) $4(4x^3 - 2x) - (3x^3 - 2x^4) = 2x^4 + 13x^3 - 8x$

7) $2(4x - 3x^3) - 3(3x^3 + 4x^2) = -15x^3 - 12x^2 + 8x$

8) $(2x^2 - 2x) - (2x^3 + 5x^2) = -2x^3 - 3x^2 - 2x$

9) $2x^3 - (4x^4 + 2x) + x^2 = -4x^4 + 2x^3 + x^2 - 2x$

10) $x^4 - 9(x^2 + x) - 5x = x^4 - 9x^2 - 14x$

11) $(-2x^2 - x^4) + (4x^4 - x^2) = 3x^4 - 3x^2$

12) $4x^2 - 5x^3 + 15x^4 - 12x^3 = 15x^4 - 17x^3 + 4x^2$

13) $2x^2 - 5x^4 + 14x^4 - 11x^3 = 9x^4 - 11x^3 + 2x^2$

14) $2x^2 + 5x^3 - 7x^2 + 12x = 5x^3 - 5x^2 + 12x$

15) $2x^4 - 5x^5 + 8x^4 - 8x^2 = -5x^5 + 10x^4 - 8x^2$

16) $5x^3 + 17x - 5x^2 - 2x^3 = 3x^3 - 5x^2 + 17x$

Evaluating One Variable

✎ *Evaluate each expression using the value given.*

1) $x = 3 \Rightarrow 6x - 9 =$

2) $x = 2 \Rightarrow 7x - 10 =$

3) $x = 1 \Rightarrow 5x + 2 =$

4) $x = 2 \Rightarrow 3x + 9 =$

5) $x = 4 \Rightarrow 4x - 8 =$

6) $x = 2 \Rightarrow 5x - 2x + 10 =$

7) $x = 3 \Rightarrow 2x - x - 6 =$

8) $x = 4 \Rightarrow 6x - 3x + 4 =$

9) $x = -2 \Rightarrow 4x - 6x - 5 =$

10) $x = -1 \Rightarrow 3x - 5x + 11 =$

11) $x = 1 \Rightarrow x - 7x + 12 =$

12) $x = 2 \Rightarrow 2(-3x + 4) =$

13) $x = 3 \Rightarrow 4(-5x - 2) =$

14) $x = 2 \Rightarrow 5(-2x - 4) =$

15) $x = -2 \Rightarrow 3(-4x - 5) =$

16) $x = 3 \Rightarrow 8x + 5 =$

17) $x = -3 \Rightarrow 12x + 9 =$

18) $x = -1 \Rightarrow 9x - 8 =$

19) $x = 2 \Rightarrow 16x - 10 =$

20) $x = 1 \Rightarrow 4x + 3 =$

21) $x = 5 \Rightarrow 7x - 2 =$

22) $x = 7 \Rightarrow 28 - x =$

23) $x = 3 \Rightarrow 5x - 10 =$

24) $x = 12 \Rightarrow 40 - 2x =$

25) $x = 2 \Rightarrow 11x - 2 =$

26) $x = 3 \Rightarrow 2x - x + 10 =$

Evaluating One Variable – Answers

✎ *Evaluate each expression using the value given.*

1) $x = 3 \Rightarrow 6x - 9 = 9$

2) $x = 2 \Rightarrow 7x - 10 = 4$

3) $x = 1 \Rightarrow 5x + 2 = 7$

4) $x = 2 \Rightarrow 3x + 9 = 15$

5) $x = 4 \Rightarrow 4x - 8 = 8$

6) $x = 2 \Rightarrow 5x - 2x + 10 = 16$

7) $x = 3 \Rightarrow 2x - x - 6 = -3$

8) $x = 4 \Rightarrow 6x - 3x + 4 = 16$

9) $x = -2 \Rightarrow 4x - 6x - 5 = -1$

10) $x = -1 \Rightarrow 3x - 5x + 11 = 13$

11) $x = 1 \Rightarrow x - 7x + 12 = 6$

12) $x = 2 \Rightarrow 2(-3x + 4) = -4$

13) $x = 3 \Rightarrow 4(-5x - 2) = -68$

14) $x = 2 \Rightarrow 5(-2x - 4) = -40$

15) $x = -2 \Rightarrow 3(-4x - 5) = 9$

16) $x = 3 \Rightarrow 8x + 5 = 29$

17) $x = -3 \Rightarrow 12x + 9 = -27$

18) $x = -1 \Rightarrow 9x - 8 = -17$

19) $x = 2 \Rightarrow 16x - 10 = 22$

20) $x = 1 \Rightarrow 4x + 3 = 7$

21) $x = 5 \Rightarrow 7x - 2 = 33$

22) $x = 7 \Rightarrow 28 - x = 21$

23) $x = 3 \Rightarrow 5x - 10 = 5$

24) $x = 12 \Rightarrow 40 - 2x = 16$

25) $x = 2 \Rightarrow 11x - 2 = 20$

26) $x = 3 \Rightarrow 2x - x + 10 = 13$

Evaluating Two Variables

✎ *Evaluate each expression using the values given.*

1) $2x + 3y, x = 2, y = 3$

_____

2) $3x + 4y, x = -1, y = -2$

_____

3) $x + 6y, x = 3, y = 1$

_____

4) $2a - (15 - b), a = 2, b = 3$

_____

5) $4a - (6 - 3b), a = 1, b = 4$

_____

6) $a - (8 - 2b), a = 2, b = 5$

_____

7) $3z + 21 + 5k, z = 4, k = 1$

_____

8) $-7a + 4b, a = 6, b = 3$

_____

9) $-4a + 3b, a = 2, b = 4$

_____

10) $-6a + 6b, a = 4, b = 3$

_____

11) $-8a + 2b, a = 4, b = 6$

_____

12) $4x + 6y, x = 6, y = 3$

_____

13) $2x + 9y, x = 8, y = 1$

_____

14) $x - 7y, x = 9, y = 4$

_____

15) $5x - 4y, x = 6, y = 3$

_____

16) $2z + 14 + 8k, z = 4, k = 1$

_____

17) $6x + 3y, x = 3, y = 8$

_____

18) $5a - 6b, a = -3, b = -1$

_____

19) $8a + 4b, a = -4, b = 3$

_____

20) $-2a - b, a = 4, b = 9$

_____

21) $-7a + 3b, a = 4, b = 3$

_____

22) $-5a + 9b, a = 7, b = 1$

_____

bit.ly/2JfrzWJ

Find more at

Evaluating Two Variables - Answers

 *Evaluate each expression using the values given.*

1) $2x + 3y, x = 2, y = 3$

    13

2) $3x + 4y, x = -1, y = -2$

    -11

3) $x + 6y, x = 3, y = 1$

    9

4) $2a - (15 - b), a = 2, b = 3$

    -8

5) $4a - (6 - 3b), a = 1, b = 4$

    10

6) $a - (8 - 2b), a = 2, b = 5$

    4

7) $3z + 21 + 5k, z = 4, k = 1$

    38

8) $-7a + 4b, a = 6, b = 3$

    -30

9) $-4a + 3b, a = 2, b = 4$

    4

10) $-6a + 6b, a = 4, b = 3$

    -6

11) $-8a + 2b, a = 4, b = 6$

    -20

12) $4x + 6y, x = 6, y = 3$

    42

13) $2x + 9y, x = 8, y = 1$

    25

14) $x - 7y, x = 9, y = 4$

    -19

15) $5x - 4y, x = 6, y = 3$

    18

16) $2z + 14 + 8k, z = 4, k = 1$

    30

17) $6x + 3y, x = 3, y = 8$

    42

18) $5a - 6b, a = -3, b = -1$

    -9

19) $8a + 4b, a = -4, b = 3$

    -20

20) $-2a - b, a = 4, b = 9$

    -17

21) $-7a + 3b, a = 4, b = 3$

    -19

22) $-5a + 9b, a = 7, b = 1$

    -26

The Distributive Property

✍ *Use the distributive property to simplify each expression.*

1) $(-3)(12x + 3) =$

2) $(-4x + 5)(-6) =$

3) $13(-4x + 2) =$

4) $7(6 - 3x) =$

5) $(6 - 5x)(-4) =$

6) $9(8 - 2x) =$

7) $(-4x + 6)5 =$

8) $(-2x + 7)(-8) =$

9) $8(-4x + 7) =$

10) $(-9x + 5)(-3) =$

11) $8(-x + 9) =$

12) $7(2 - 6x) =$

13) $(-12x + 4)(-3) =$

14) $(-6)(-10x + 6) =$

15) $(-5)(5 - 11x) =$

16) $9(4 - 8x) =$

17) $(-6x + 2)7 =$

18) $(-9)(1 - 12x) =$

19) $(-3)(4 - 6x) =$

20) $(2 - 8x)(-2) =$

21) $20(2 - x) =$

22) $12(-4x + 3) =$

23) $15(2 - 3x) =$

24) $(-4x + 5)2 =$

25) $(-11x + 8)(-2) =$

26) $14(5 - 8x) =$

The Distributive Property - Answers

✎ *Use the distributive property to simplify each expression.*

1) $(-3)(12x + 3) = -36x - 9$

2) $(-4x + 5)(-6) = 24x - 30$

3) $13(-4x + 2) = -52x + 26$

4) $7(6 - 3x) = -21x + 42$

5) $(6 - 5x)(-4) = 20x - 24$

6) $9(8 - 2x) = -18x + 72$

7) $(-4x + 6)5 = -20x + 30$

8) $(-2x + 7)(-8) = 16x - 56$

9) $8(-4x + 7) = -32x + 56$

10) $(-9x + 5)(-3) = 27x - 15$

11) $8(-x + 9) = -8x + 72$

12) $7(2 - 6x) = -42x + 14$

13) $(-12x + 4)(-3) = 36x - 12$

14) $(-6)(-10x + 6) = 60x - 36$

15) $(-5)(5 - 11x) = 55x - 25$

16) $9(4 - 8x) = -72x + 36$

17) $(-6x + 2)7 = -42x + 14$

18) $(-9)(1 - 12x) = 108x - 9$

19) $(-3)(4 - 6x) = 18x - 12$

20) $(2 - 8x)(-2) = 16x - 4$

21) $20(2 - x) = -20x + 40$

22) $12(-4x + 3) = -48x + 36$

23) $15(2 - 3x) = -45x + 30$

24) $(-4x + 5)2 = -8x + 10$

25) $(-11x + 8)(-2) = 22x - 16$

26) $14(5 - 8x) = -112x + 70$

One–Step Equations

✎ *Solve each equation for* x.

1) $x - 15 = 24 \Rightarrow x = $ _____

2) $18 = -6 + x \Rightarrow x = $ ____

3) $19 - x = 8 \Rightarrow x = $ ____

4) $x - 22 = 24 \Rightarrow x = $ ____

5) $24 - x = 17 \Rightarrow x = $ ____

6) $16 - x = 3 \Rightarrow x = $ ____

7) $x + 14 = 12 \Rightarrow x = $ ____

8) $26 + x = 8 \Rightarrow x = $ ____

9) $x + 9 = -18 \Rightarrow x = $ ____

10) $x + 21 = 11 \Rightarrow x = $ ____

11) $17 = -5 + x \Rightarrow x = $ ____

12) $x + 20 = 29 \Rightarrow x = $ ____

13) $x - 13 = 19 \Rightarrow x = $ ____

14) $x + 9 = -17 \Rightarrow x = $ ____

15) $x + 4 = -23 \Rightarrow x = $ ____

16) $16 = -9 + x \Rightarrow x = $ ____

17) $4x = 28 \Rightarrow x = $ ____

18) $21 = -7x \Rightarrow x = $ ____

19) $12x = -12 \Rightarrow x = $ ____

20) $13x = 39 \Rightarrow x = $ ____

21) $8x = -16 \Rightarrow x = $ ____

22) $\frac{x}{2} = -5 \Rightarrow x = $ ____

23) $\frac{x}{9} = 6 \Rightarrow x = $ ____

24) $27 = \frac{x}{5} \Rightarrow x = $ ____

25) $\frac{x}{4} = -3 \Rightarrow x = $ ____

26) $x \div 8 = 7 \Rightarrow x = $ ____

27) $x \div 2 = -3 \Rightarrow x = $ ____

28) $4x = 48 \Rightarrow x = $ ____

29) $9x = 72 \Rightarrow x = $ ____

30) $8x = -32 \Rightarrow x = $ ____

31) $80 = -10x \Rightarrow x = $ ____

One – Step Equations - Answers

✍ *Solve each equation for x.*

1) $x - 15 = 24 \Rightarrow x = 39$

2) $18 = -6 + x \Rightarrow x = 24$

3) $19 - x = 8 \Rightarrow x = 11$

4) $x - 22 = 24 \Rightarrow x = 46$

5) $24 - x = 17 \Rightarrow x = 7$

6) $16 - x = 3 \Rightarrow x = 13$

7) $x + 14 = 12 \Rightarrow x = 26$

8) $26 + x = 8 \Rightarrow x = -18$

9) $x + 9 = -18 \Rightarrow x = -27$

10) $x + 21 = 11 \Rightarrow x = -10$

11) $17 = -5 + x \Rightarrow x = 22$

12) $x + 20 = 29 \Rightarrow x = 9$

13) $x - 13 = 19 \Rightarrow x = 32$

14) $x + 9 = -17 \Rightarrow x = -26$

15) $x + 4 = -23 \Rightarrow x = -27$

16) $16 = -9 + x \Rightarrow x = 25$

17) $4x = 28 \Rightarrow x = 7$

18) $21 = -7x \Rightarrow x = -3$

19) $12x = -12 \Rightarrow x = -1$

20) $13x = 39 \Rightarrow x = 3$

21) $8x = -16 \Rightarrow x = -2$

22) $\frac{x}{2} = -5 \Rightarrow x = -10$

23) $\frac{x}{9} = 6 \Rightarrow x = 54$

24) $27 = \frac{x}{5} \Rightarrow x = 135$

25) $\frac{x}{4} = -3 \Rightarrow x = -12$

26) $x \div 8 = 7 \Rightarrow x = 56$

27) $x \div 2 = -3 \Rightarrow x = -6$

28) $4x = 48 \Rightarrow x = 12$

29) $9x = 72 \Rightarrow x = 8$

30) $8x = -32 \Rightarrow x = -4$

31) $80 = -10x \Rightarrow x = -8$

Multi – Step Equations

✍ *Solve each equation.*

1) $3x - 8 = 13 \Rightarrow x = $ ____

2) $23 = - (x - 5) \Rightarrow x = $ ____

3) $-(8 - x) = 15 \Rightarrow x = $ ____

4) $29 = -x + 12 \Rightarrow x = $ ____

5) $2(3 - 2x) = 10 \Rightarrow x = $ ____

6) $3x - 3 = 15 \Rightarrow x = $ ____

7) $32 = -x + 15 \Rightarrow x = $ ____

8) $-(10 - x) = -13 \Rightarrow x = $ ____

9) $-4(7 + x) = 4 \Rightarrow x = $ ____

10) $23 = 2x - 8 \Rightarrow x = $ ____

11) $-6(3 + x) = 6 \Rightarrow x = $ ____

12) $-3 = 3x - 15 \Rightarrow x = $ ____

13) $-7(12 + x) = 7 \Rightarrow x = $ ____

14) $8(6 - 4x) = 16 \Rightarrow x = $ ____

15) $18 - 4x = -9 - x \Rightarrow x = $ ____

16) $6(4 - x) = 30 \Rightarrow x = $ ____

17) $15 - 3x = -5 - x \Rightarrow x = $ ____

18) $9(-7 - 3x) = 18 \Rightarrow x = $ ____

19) $16 - 2x = -4 - 7x \Rightarrow x = $ ____

20) $14 - 2x = 14 + x \Rightarrow x = $ ____

21) $21 - 3x = -7 - 10x \Rightarrow x = $ ___

22) $8 - 2x = 11 + x \Rightarrow x = $ ____

23) $10 + 12x = -8 + 6x \Rightarrow x = $ ____

24) $25 + 20x = -5 + 5x \Rightarrow x = $ ___

25) $16 - x = -8 - 7x \Rightarrow x = $ ____

26) $17 - 3x = 13 + x \Rightarrow x = $ ____

27) $22 + 5x = -8 - x \Rightarrow x = $ ____

28) $-9(7 + x) = 9 \Rightarrow x = $ ___

29) $11 + 3x = -4 - 2x \Rightarrow x = $ ___

30) $13 - 2x = 3 - 3x \Rightarrow x = $ ___

31) $19 - x = -1 - 11x \Rightarrow x = $ ____

32) $12 - 2x = -2 - 4x \Rightarrow x = $ ____

Multi –Step Equations - Answers

✍ *Solve each equation.*

1) $3x - 8 = 13 \Rightarrow x = 7$

2) $23 = -(x - 5) \Rightarrow x = -18$

3) $-(8 - x) = 15 \Rightarrow x = 23$

4) $29 = -x + 12 \Rightarrow x = -17$

5) $2(3 - 2x) = 10 \Rightarrow x = -1$

6) $3x - 3 = 15 \Rightarrow x = 6$

7) $32 = -x + 15 \Rightarrow x = -17$

8) $-(10 - x) = -13 \Rightarrow x = -3$

9) $-4(7 + x) = 4 \Rightarrow x = -8$

10) $23 = 2x - 8 \Rightarrow x = 15.5$

11) $-6(3 + x) = 6 \Rightarrow x = -4$

12) $-3 = 3x - 15 \Rightarrow x = 4$

13) $-7(12 + x) = 7 \Rightarrow x = -13$

14) $8(6 - 4x) = 16 \Rightarrow x = 1$

15) $18 - 4x = -9 - x \Rightarrow x = 9$

16) $6(4 - x) = 30 \Rightarrow x = -1$

17) $15 - 3x = -5 - x \Rightarrow x = 10$

18) $9(-7 - 3x) = 18 \Rightarrow x = -3$

19) $16 - 2x = -4 - 7x \Rightarrow x = -4$

20) $14 - 2x = 14 + x \Rightarrow x = 0$

21) $21 - 3x = -7 - 10x \Rightarrow x = -4$

22) $8 - 2x = 11 + x \Rightarrow x = -1$

23) $10 + 12x = -8 + 6x \Rightarrow x = -3$

24) $25 + 20x = -5 + 5x \Rightarrow x = -2$

25) $16 - x = -8 - 7x \Rightarrow x = -4$

26) $17 - 3x = 13 + x \Rightarrow x = 1$

27) $22 + 5x = -8 - x \Rightarrow x = -5$

28) $-9(7 + x) = 9 \Rightarrow x = -8$

29) $11 + 3x = -4 - 2x \Rightarrow x = -3$

30) $13 - 2x = 3 - 3x \Rightarrow x = -10$

31) $19 - x = -1 - 11x \Rightarrow x = -2$

32) $12 - 2x = -2 - 4x \Rightarrow x = -7$

Graphing Single–Variable Inequalities

✎ *Graph each inequality*

1) $x < 6$

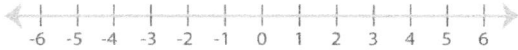

2) $x \geq 1$

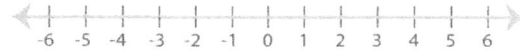

3) $x \geq -6$

4) $x \leq -2$

5) $x > -1$

6) $3 > x$

7) $2 \leq x$

8) $x > 0$

9) $-3 \leq x$

10) $-4 \leq x$

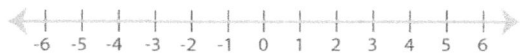

11) $x \leq 5$

12) $0 \leq x$

13) $-5 \leq x$

14) $x > -6$

EffortlessMath.com

Find more at

bit.ly/3aJ4GGo

Graphing Single–Variable Inequalities - Answers

✎ *Graph each inequality.*

1) $x < 6$

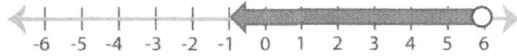

2) $x \geq 1$

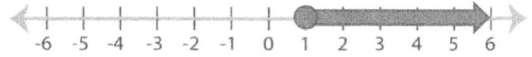

3) $x \geq -6$

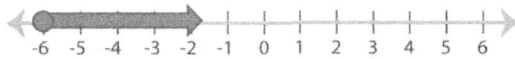

4) $x \leq -2$

5) $x > -1$

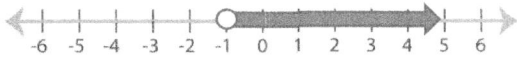

6) $3 > x$

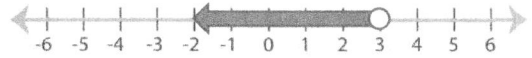

7) $2 \leq x$

8) $x > 0$

9) $-3 \leq x$

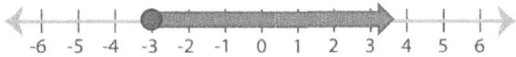

10) $-4 \leq x$

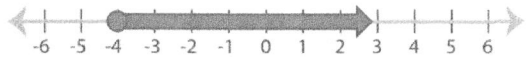

11) $x \leq 5$

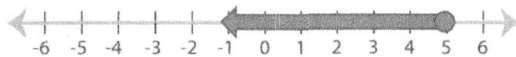

12) $0 \leq x$

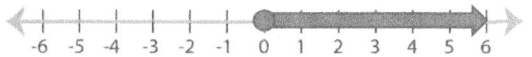

13) $-5 \leq x$

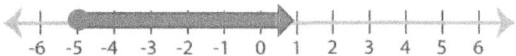

14) $x > -6$

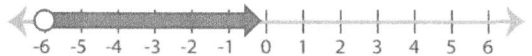

One–Step Inequalities

✎ Solve each inequality for x.

1) $x - 10 < 22 \Rightarrow$ _____

2) $18 \leq -4 + x \Rightarrow$ _____

3) $x - 33 > 8 \Rightarrow$ _____

4) $x + 22 \geq 24 \Rightarrow$ _____

5) $x - 24 > 17 \Rightarrow$ _____

6) $x + 5 \geq 3 \Rightarrow x$_____

7) $x + 14 < 12 \Rightarrow$ _____

8) $26 + x \leq 8 \Rightarrow$ _____

9) $x + 9 \geq -18 \Rightarrow$ _____

10) $x + 24 < 11 \Rightarrow$ _____

11) $17 \leq -5 + x \Rightarrow$ _____

12) $x + 25 > 29 \Rightarrow x$_____

13) $x - 17 \geq 19 \Rightarrow$ _____

14) $x + 8 > -17 \Rightarrow$ _____

15) $x + 8 < -23 \Rightarrow$ _____

16) $16 \leq -5 + x \Rightarrow$ _____

17) $4x \leq 12 \Rightarrow$ _____

18) $28 \geq -7x \Rightarrow$ _____

19) $2x > -14 \Rightarrow$ _____

20) $13x \leq 39 \Rightarrow$ _____

21) $-8x > -16 \Rightarrow$ _____

22) $\frac{x}{2} < -6 \Rightarrow$ _____

23) $\frac{x}{6} > 6 \Rightarrow$ _____

24) $27 \leq \frac{x}{4} \Rightarrow$ _____

25) $\frac{x}{8} < -3 \Rightarrow$ _____

26) $6x \geq 18 \Rightarrow$ _____

27) $5x \geq -25 \Rightarrow$ _____

28) $4x > 48 \Rightarrow$ _____

29) $8x \leq 72 \Rightarrow$ _____

30) $-4x < -32 \Rightarrow$ _____

31) $40 > -10x \Rightarrow$ _____

One–Step Inequalities - Answers

✎ *Solve each inequality for x.*

1) $x - 10 < 22 \Rightarrow x < 32$

2) $18 \le -4 + x \Rightarrow 22 \le x$

3) $x - 33 > 8 \Rightarrow x > 41$

4) $x + 22 \ge 24 \Rightarrow x \ge 2$

5) $x - 24 > 17 \Rightarrow x > 41$

6) $x + 5 \ge 3 \Rightarrow x \ge -2$

7) $x + 14 < 12 \Rightarrow x < -2$

8) $26 + x \le 8 \Rightarrow x \le -18$

9) $x + 9 \ge -18 \Rightarrow x \ge -27$

10) $x + 24 < 11 \Rightarrow x < -13$

11) $17 \le -5 + x \Rightarrow 22 \le x$

12) $x + 25 > 29 \Rightarrow x > 4$

13) $x - 17 \ge 19 \Rightarrow x \ge 36$

14) $x + 8 > -17 \Rightarrow x > -25$

15) $x + 8 < -23 \Rightarrow x < -31$

16) $16 \le -5 + x \Rightarrow 21 \le x$

17) $4x \le 12 \Rightarrow x \le 3$

18) $28 \ge -7x \Rightarrow -4 \le x$

19) $2x > -14 \Rightarrow x > -7$

20) $13x \le 39 \Rightarrow x \le 3$

21) $-8x > -16 \Rightarrow x < 2$

22) $\frac{x}{2} < -6 \Rightarrow x < -12$

23) $\frac{x}{6} > 6 \Rightarrow x > 36$

24) $27 \le \frac{x}{4} \Rightarrow 108 \le x$

25) $\frac{x}{8} < -3 \Rightarrow x < -24$

26) $6x \ge 18 \Rightarrow x \ge 3$

27) $5x \ge -25 \Rightarrow x \ge -5$

28) $4x > 48 \Rightarrow x > 12$

29) $8x \le 72 \Rightarrow x \le 9$

30) $-4x < -32 \Rightarrow x > 8$

31) $40 > -10x \Rightarrow -4 < x$

Multi –Step Inequalities

✎ **Solve each inequality.**

1) $2x - 8 \leq 8 \rightarrow$ _____

2) $3 + 2x \geq 17 \rightarrow$ _____

3) $5 + 3x \geq 26 \rightarrow$ _____

4) $2x - 8 \leq 14 \rightarrow$ _____

5) $3x - 4 \leq 23 \rightarrow$ _____

6) $7x - 5 \leq 51 \rightarrow$ _____

7) $4x - 9 \leq 27 \rightarrow$ _____

8) $6x - 11 \leq 13 \rightarrow$ _____

9) $5x - 7 \leq 33 \rightarrow$ _____

10) $6 + 2x \geq 28 \rightarrow$ _____

11) $8 + 3x \geq 35 \rightarrow$ _____

12) $4 + 6x < 34 \rightarrow$ _____

13) $3 + 2x \geq 53 \rightarrow$ _____

14) $7 - 6x > 56 + x \rightarrow$ _____

15) $9 + 4x \geq 39 + 2x \rightarrow$ _____

16) $3 + 5x \geq 43 \rightarrow$ _____

17) $4 - 7x < 60 \rightarrow$ _____

18) $11 - 4x \geq 55 \rightarrow$ _____

19) $12 + x \geq 48 - 2x \rightarrow$ _____

20) $10 - 10x \leq -20 \rightarrow$ _____

21) $5 - 9x \geq -40 \rightarrow$ _____

22) $8 - 7x \geq 36 \rightarrow$ _____

23) $5 + 11x < 69 + 3x \rightarrow$ _____

24) $6 + 8x < 28 - 3x \rightarrow$ _____

25) $9 + 11x < 57 - x \rightarrow$ _____

26) $3 + 10x \geq 45 - 4x \rightarrow$ _____

Multi –Step Inequalities - Answers

✍ *Solve each inequality.*

1) $2x - 8 \leq 8 \to x \leq 8$

2) $3 + 2x \geq 17 \to x \geq 7$

3) $5 + 3x \geq 26 \to x \geq 7$

4) $2x - 8 \leq 14 \to x \leq 11$

5) $3x - 4 \leq 23 \to x \leq 9$

6) $7x - 5 \leq 51 \to x \leq 8$

7) $4x - 9 \leq 27 \to x \leq 9$

8) $6x - 11 \leq 13 \to x \leq 4$

9) $5x - 7 \leq 33 \to x \leq 8$

10) $6 + 2x \geq 28 \to x \geq 11$

11) $8 + 3x \geq 35 \to x \geq 9$

12) $4 + 6x < 34 \to x < 5$

13) $3 + 2x \geq 53 \to x \geq 25$

14) $7 - 6x > 56 + x \to x < -7$

15) $9 + 4x \geq 39 + 2x \to x \geq 15$

16) $3 + 5x \geq 43 \to x \geq 8$

17) $4 - 7x < 60 \to x > -8$

18) $11 - 4x \geq 55 \to x \leq -11$

19) $12 + x \geq 48 - 2x \to x \geq 12$

20) $10 - 10x \leq -20 \to x \geq 3$

21) $5 - 9x \geq -40 \to x \leq 5$

22) $8 - 7x \geq 36 \to x \leq -4$

23) $5 + 11x < 69 + 3x \to x < 8$

24) $6 + 8x < 28 - 3x \to x < 2$

25) $9 + 11x < 57 - x \to x < 4$

26) $3 + 10x \geq 45 - 4x \to x \geq 3$

System of Equations

✎ *Solve each system of equations.*

1) $-x + y = 2$
   $-2x + y = 3$
   $x =$
   $y =$

2) $-5x + y = -3$
   $3x - 8y = 24$
   $x =$
   $y =$

3) $y = -5$
   $4x - 5y = 13$
   $x =$

4) $y = -6x + 8$
   $5x - 4y = -3$
   $x =$
   $y =$

5) $10x - 8y = -15$
   $-6x + 4y = 13$
   $x =$
   $y =$

6) $-3x - 4y = 5$
   $x - 2y = 5$
   $x =$
   $y =$

7) $5x - 12y = -19$
   $-6x + 7y = 8$
   $x =$
   $y =$

8) $5x - 7y = -2$
   $-x - 2y = -3$
   $x =$
   $y =$

9) $-x + 3y = 3$
   $-7x + 8y = -5$
   $x =$
   $y =$

10) $-4x + 3y = -18$
    $4x - y = 14$
    $x =$
    $y =$

11) $6x - 7y = -8$
    $-x - 4y = -9$
    $x =$
    $y =$

12) $-3x + 2y = -16$
    $4x - y = 13$
    $x =$
    $y =$

System of Equations- Answers

✎ *Solve each system of equations.*

1) $-x + y = 2$
   $-2x + y = 3$
   $x = -1$
   $y = 1$

2) $-5x + y = -3$
   $3x - 8y = 24$
   $x = 0$
   $y = -3$

3) $y = -5$
   $4x - 5y = 13$
   $x = -3$

4) $y = -6x + 8$
   $5x - 4y = -3$
   $x = 1$
   $y = 2$

5) $10x - 8y = -15$
   $-6x + 4y = 13$
   $x = -\dfrac{11}{2}$
   $y = -5$

6) $-3x - 4y = 5$
   $x - 2y = 5$
   $x = 1$
   $y = -2$

7) $5x - 12y = -19$
   $-6x + 7y = 8$
   $x = 1$
   $y = 2$

8) $5x - 7y = -2$
   $-x - 3y = -3$
   $x = 1$
   $y = 1$

9) $-x + 3y = 3$
   $-7x + 8y = -5$
   $x = 3$
   $y = 2$

10) $-4x + 3y = -18$
    $4x - y = 14$
    $x = 3$
    $y = -2$

11) $6x - 7y = -8$
    $-x - 4y = -9$
    $x = 1$
    $y = 2$

12) $-3x + 2y = -16$
    $4x - y = 13$
    $x = 2$
    $y = -5$

Finding Slope

✎ *Find the slope of each line.*

1) $y = x - 5$, Slope $=$

2) $y = -3x + 2$, Slope $=$

3) $y = -x - 1$, Slope $=$

4) $y = -x - 9$, Slope $=$

5) $y = 5 + 2x$, Slope $=$

6) $y = 1 - 8x$, Slope $=$

7) $y = -4x + 3$, Slope $=$

8) $y = -9x + 8$, Slope $=$

9) $y = -2x + 4$, Slope $=$

10) $y = 9x - 8$, Slope $=$

11) $y = \frac{1}{2}x + 4$, Slope $=$

12) $y = -\frac{2}{5}x + 7$, Slope $=$

13) $-x + 3y = 5$, Slope $=$

14) $4x + 4y = 6$, Slope $=$

15) $6y - 2x = 10$, Slope $=$

16) $7y - x = 2$, Slope $=$

✎ *Find the slope of the line through each pair of points.*

1) $(4, 4), (8, 12)$, Slope $=$

2) $(-2, 4), (0, 6)$, Slope $=$

3) $(6, -2), (2, 6)$, Slope $=$

4) $(-4, -2), (0, 6)$, Slope $=$

5) $(6, 2), (3, 5)$, Slope $=$

6) $(-5, 1), (-1, 9)$, Slope $=$

7) $(8, 4), (9, 6)$, Slope $=$

8) $(10, -1), (7, 8)$, Slope $=$

9) $(14, -7), (13, -6)$, Slope $=$

10) $(10, 7), (8, 1)$, Slope $=$

11) $(5, 1), (8, 10)$, Slope $=$

12) $(9, -10), (8, 12)$, Slope $=$

Finding Slope - answer

✎ *Find the slope of each line.*

1) $y = x - 5$, Slope $= 1$

2) $y = -3x + 2$, Slope $= -3$

3) $y = -x - 1$, Slope $= -1$

4) $y = -x - 9$, Slope $= -1$

5) $y = 5 + 2x$, Slope $= 2$

6) $y = 1 - 8x$, Slope $= -8$

7) $y = -4x + 3$, Slope $= -4$

8) $y = -9x + 8$, Slope $= -9$

9) $y = -2x + 4$, Slope $= -2$

10) $y = 9x - 8$, Slope $= 9$

11) $y = \frac{1}{2}x + 4$, Slope $= 0.5$

12) $y = -\frac{2}{5}x + 7$, Slope $= -0.4$

13) $-x + 3y = 5$, Slope $= \frac{1}{3}$

14) $4x + 4y = 6$, Slope $= -1$

15) $6y - 2x = 10$, Slope $= \frac{1}{3}$

16) $7y - x = 2$, Slope $= \frac{1}{7}$

✎ *Find the slope of the line through each pair of points.*

1) $(4, 4), (8, 12)$, Slope $= 2$

2) $(-2, 4), (0, 6)$, Slope $= 1$

3) $(6, -2), (2, 6)$, Slope $= -2$

4) $(-4, -2), (0, 6)$, Slope $= 2$

5) $(6, 2), (3, 5)$, Slope $= -1$

6) $(-5, 1), (-1, 9)$, Slope $= 2$

7) $(8, 4), (9, 6)$, Slope $= 2$

8) $(10, -1), (7, 8)$, Slope $= -3$

9) $(14, -7), (13, -6)$, Slope $= -1$

10) $(10, 7), (8, 1)$, Slope $= 3$

11) $(5, 1), (8, 10)$, Slope $= 3$

12) $(9, -10), (8, 12)$, Slope $= -22$

Graphing Lines Using Slope–Intercept Form

✎ *Sketch the graph of each line.*

1) $y = -x + 1$

2) $y = 2x - 3$

3) $y = -x + 2$

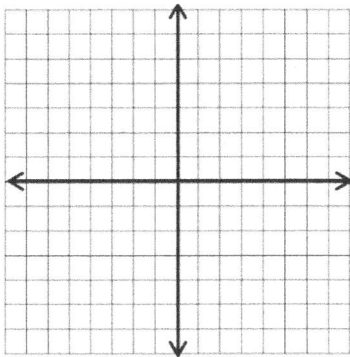
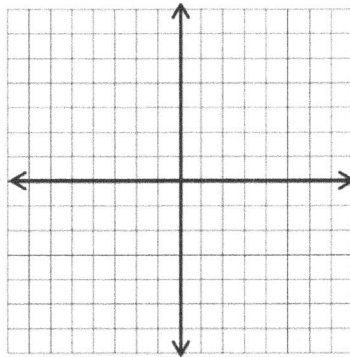
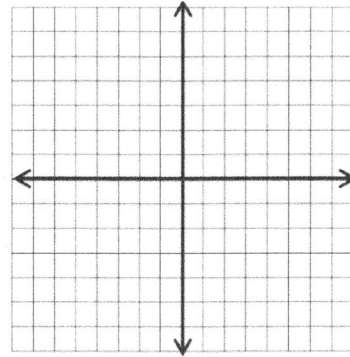

4) $y = x + 1$

5) $y = 2x - 4$

6) $y = -\frac{1}{2}x + 1$

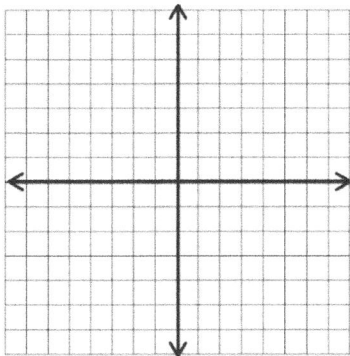
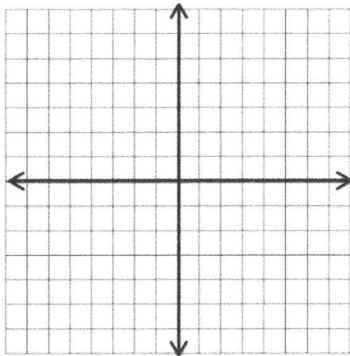
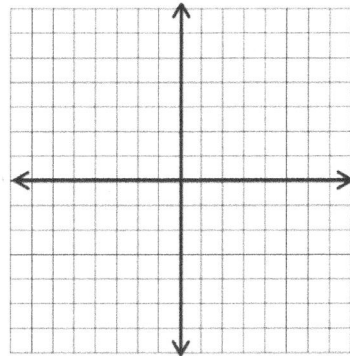

bit.ly/3hfdnJL

Find more at

Graphing Lines Using Slope–Intercept Form - Answers

✎ *Sketch the graph of each line.*

1) $y = -x + 1$

2) $y = 2x - 3$

3) $y = -x + 2$

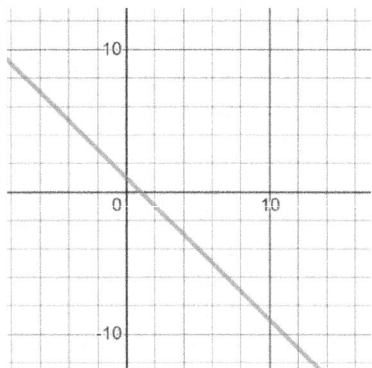

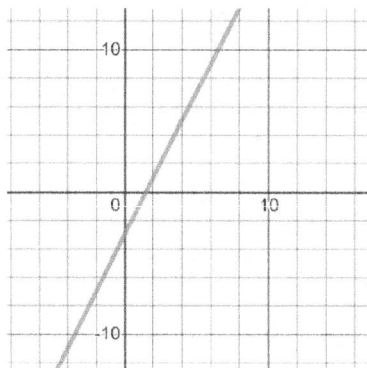

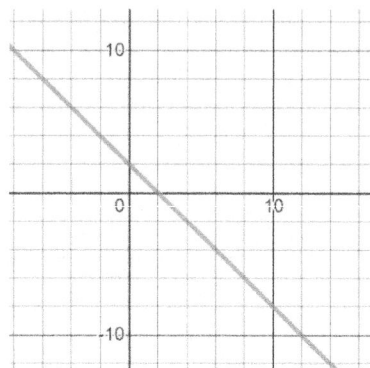

4) $y = x + 1$

5) $y = 2x - 4$

6) $y = -\frac{1}{2}x + 1$

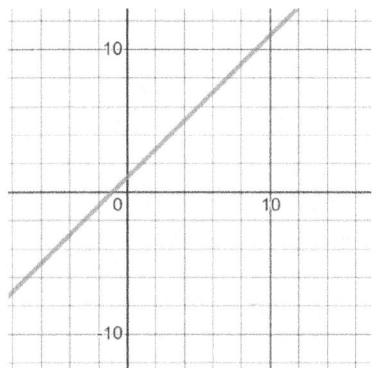

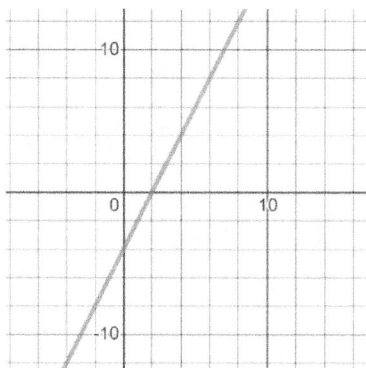

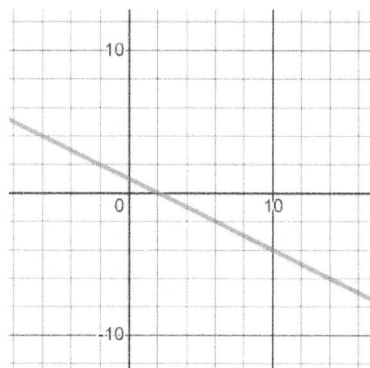

Writing Linear Equations

✎ *Write the equation of the line through the given points.*

1) through: $(1, -2), (2, 4)$

$y =$

2) through: $(-2, 3), (1, 6)$

$y =$

3) through: $(-1, 2), (3, 6)$

$y =$

4) through: $(8, 5), (5, 2)$

$y =$

5) through: $(7, -10), (2, 10)$

$y =$

6) through: $(7, 2), (6, 1)$

$y =$

7) through: $(6, -1), (4, 1)$

$y =$

8) through: $(-2, 8), (-4, -6)$

$y =$

9) through: $(-2, 5), (-3, 4)$

$y =$

10) through: $(6, 8), (8, -6)$

$y =$

11) through: $(-2, 5), (-4, -3)$

$y =$

12) through: $(8, 8), (4, -8)$

$y =$

13) through: $(7, -4)$, Slope: -1

$y =$

14) through: $(4, -10)$, Slope: -2

$y =$

15) through: $(6, 10)$, Slope: 9

$y =$

16) through: $(-6, 8)$, Slope: -2

$y =$

✎ *Solve each problem.*

1) What is the equation of a line with slope 8 and intercept 5? _____

2) What is the equation of a line with slope 4 and intercept 10? _____

3) What is the equation of a line with slope 9 and passes through point $(5, 23)$? _____

4) What is the equation of a line with slope -7 and passes through point $(-3, 18)$? _____

Writing Linear Equations - Answers

✍ **Write the equation of the line through the given points.**

1) through: $(1, -2), (2, 4)$

$$y = 6x - 8$$

2) through: $(-2, 3), (1, 6)$

$$y = x + 5$$

3) through: $(-1, 2), (3, 6)$

$$y = x + 3$$

4) through: $(8, 5), (5, 2)$

$$y = x - 3$$

5) through: $(7, -10), (2, 10)$

$$y = -4x + 18$$

6) through: $(7, 2), (6, 1)$

$$y = x - 5$$

7) through: $(6, -1), (4, 1)$

$$y = -x + 5$$

8) through: $(-2, 8), (-4, -6)$

$$y = 7x + 22$$

9) through: $(-2, 5), (-3, 4)$

$$y = x + 7$$

10) through: $(6, 8), (8, -6)$

$$y = -7x + 50$$

11) through: $(-2, 5), (-4, -3)$

$$y = 4x + 13$$

12) through: $(8, 8), (4, -8)$

$$y = 4x - 24$$

13) through: $(7, -4)$, Slope: -1

$$y = -x + 3$$

14) through: $(4, -10)$, Slope: -2

$$y = -2x - 2$$

15) through: $(6, 10)$, Slope: 9

$$y = 9x - 44$$

16) through: $(-6, 8)$, Slope: -2

$$y = -2x - 4$$

✍ **Solve each problem.**

1) What is the equation of a line with slope 8 and intercept 5? $y = 8x + 5$

2) What is the equation of a line with slope 4 and intercept 10? $y = 4x + 10$

3) What is the equation of a line with slope 9 and passes through point $(5, 23)$? $y = 9x - 22$

4) What is the equation of a line with slope -7 and passes through point $(-3, 18)$? $y = -7x - 3$

Finding Midpoint

Find the midpoint of the line segment with the given endpoints.

1) $(2, 2), (0, 4),$

   $midpoint = (\_\_, \_\_)$

2) $(3, 3), (-1, 5),$

   $midpoint = (\_\_, \_\_)$

3) $(2, -1), (0, 5),$

   $midpoint = (\_\_, \_\_)$

4) $(-3, 7), (-1, 5),$

   $midpoint = (\_\_, \_\_)$

5) $(5, -2), (9, -6),$

   $midpoint = (\_\_, \_\_)$

6) $(-6, -3), (4, -7),$

   $midpoint = (\_\_, \_\_)$

7) $(7, 0), (-7, 8),$

   $midpoint = (\_\_, \_\_)$

8) $(-8, 4), (-4, 0),$

   $midpoint = (\_\_, \_\_)$

9) $(-3, 6), (9, -8),$

   $midpoint = (\_\_, \_\_)$

10) $(6, 8), (6, -6),$

    $midpoint = (\_\_, \_\_)$

11) $(6, 7), (-8, 5),$

    $midpoint = (\_\_, \_\_)$

12) $(9, 3), (-3, -9),$

    $midpoint = (\_\_, \_\_)$

13) $(-6, 12), (-4, 6),$

    $midpoint = (\_\_, \_\_)$

14) $(10, 7), (8, -3),$

    $midpoint = (\_\_, \_\_)$

15) $(13, 7), (-5, 3),$

    $midpoint = (\_\_, \_\_)$

16) $(-9, -4), (-5, 8),$

    $midpoint = (\_\_, \_\_)$

17) $(11, 7), (5, 13),$

    $midpoint = (\_\_, \_\_)$

18) $(-7, -10), (11, -2),$

    $midpoint = (\_\_, \_\_)$

19) $(10, 15), (-4, 9),$

    $midpoint = (\_\_, \_\_)$

20) $(11, -4), (7, 12),$

    $midpoint = (\_\_, \_\_)$

Finding Midpoint - Answers

✍ *Find the midpoint of the line segment with the given endpoints.*

1) $(2, 2), (0, 4),$
   $midpoint = (1, 3)$

2) $(3, 3), (-1, 5),$
   $midpoint = (1, 4)$

3) $(2, -1), (0, 5),$
   $midpoint = (1, 2)$

4) $(-3, 7), (-1, 5),$
   $midpoint = (-2, 6)$

5) $(5, -2), (9, -6),$
   $midpoint = (7, -4)$

6) $(-6, -3), (4, -7),$
   $midpoint = (-1, -5)$

7) $(7, 0), (-7, 8),$
   $midpoint = (0, 4)$

8) $(-8, 4), (-4, 0),$
   $midpoint = (-6, 2)$

9) $(-3, 6), (9, -8),$
   $midpoint = (3, -1)$

10) $(6, 8), (6, -6),$
   $midpoint = (6, 1)$

11) $(6, 7), (-8, 5),$
   $midpoint = (-1, 6)$

12) $(9, 3), (-3, -9),$
   $midpoint = (3, -3)$

13) $(-6, 12), (-4, 6),$
   $midpoint = (-5, 9)$

14) $(10, 7), (8, -3),$
   $midpoint = (9, 2)$

15) $(13, 7), (-5, 3),$
   $midpoint = (4, 5)$

16) $(-9, -4), (-5, 8),$
   $midpoint = (-7, 2)$

17) $(11, 7), (5, 13),$
   $midpoint = (8, 10)$

18) $(-7, -10), (11, -2),$
   $midpoint = (2, -6)$

19) $(10, 15), (-4, 9),$
   $midpoint = (3, 12)$

20) $(11, -4), (7, 12),$
   $midpoint = (9, 4)$

Finding Distance of Two Points

✎ *Find the distance of each pair of points.*

1) $(1, 9), (5, 6),$

   Distance = ____

2) $(-4, 5), (8, 10),$

   Distance = ____

3) $(5, -2), (-3, 4),$

   Distance = ____

4) $(-3, 0), (3, 8),$

   Distance = ____

5) $(-5, 3), (4, -9),$

   Distance = ____

6) $(-7, -5), (5, 0),$

   Distance = ____

7) $(4, 3), (-4, -12),$

   Distance = ____

8) $(10, 1), (-5, -19),$

   Distance = ____

9) $(3, 3), (-1, 5),$

   Distance = ____

10) $(2, -1), (10, 5),$

    Distance = ____

11) $(-3, 7), (-1, 4),$

    Distance = ____

12) $(5, -2), (9, -5),$

    Distance = ____

13) $(-8, 4), (4, 9),$

    Distance = ____

14) $(6, 8), (6, -6),$

    Distance = ____

15) $(9, 3), (-3, -2),$

    Distance = ____

16) $(-4, 12), (-4, 6),$

    Distance = ____

17) $(-9, -4), (-4, 8),$

    Distance = ____

18) $(11, 7), (3, 22),$

    Distance = ____

Finding Distance of Two Points - Answers

✎ *Find the distance of each pair of points.*

1) $(1,9),(5,6)$,

  Distance $= 5$

2) $(-4,5),(8,10)$,

  Distance $= 13$

3) $(5,-2),(-3,4)$,

  Distance $= 10$

4) $(-3,0),(3,8)$,

  Distance $= 10$

5) $(-5,3),(4,-9)$,

  Distance $= 15$

6) $(-7,-5),(5,0)$,

  Distance $= 13$

7) $(4,3),(-4,-12)$,

  Distance $= 17$

8) $(10,1),(-5,-19)$,

  Distance $= 25$

9) $(3,3),(-1,5)$,

  Distance $= \sqrt{20} = 2\sqrt{5}$

10) $(2,-1),(10,5)$,

  Distance $= 10$

11) $(-3,7),(-1,4)$,

  Distance $= \sqrt{13}$

12) $(5,-2),(9,-5)$,

  Distance $= 5$

13) $(-8,4),(4,9)$,

  Distance $= 13$

14) $(6,8),(6,-6)$,

  Distance $= 14$

15) $(9,3),(-3,-2)$,

  Distance $= 13$

16) $(-4,12),(-4,6)$,

  Distance $= 6$

17) $(-9,-4),(-4,8)$,

  Distance $= 13$

18) $(11,7),(3,22)$,

  Distance $= 17$

Multiplication Property of Exponents

✍ *Simplify and write the answer in exponential form.*

1) $2 \times 2^2 =$

2) $5^3 \times 5 =$

3) $3^2 \times 3^2 =$

4) $4^2 \times 4^2 =$

5) $7^3 \times 7^2 \times 7 =$

6) $2 \times 2^2 \times 2^2 =$

7) $5^3 \times 5^2 \times 5 \times 5 =$

8) $2x \times x =$

9) $x^3 \times x^2 =$

10) $x^4 \times x^4 =$

11) $x^2 \times x^2 \times x^2 =$

12) $6x \times 6x =$

13) $2x^2 \times 2x^2 =$

14) $3x^2 \times x =$

15) $4x^4 \times 4x^4 \times 4x^4 =$

16) $2x^2 \times x^2 =$

17) $x^4 \times 3x =$

18) $x \times 2x^2 =$

19) $5x^4 \times 5x^4 =$

20) $2yx^2 \times 2x =$

21) $3x^4 \times y^2x^4 =$

22) $y^2x^3 \times y^5x^2 =$

23) $4yx^3 \times 2x^2y^3 =$

24) $6x^2 \times 6x^3y^4 =$

25) $3x^4y^5 \times 7x^2y^3 =$

26) $7x^2y^5 \times 9xy^3 =$

27) $7xy^4 \times 4x^3y^3 =$

28) $3x^5y^3 \times 8x^2y^3 =$

29) $3x \times y^5x^3 \times y^4 =$

30) $yx^2 \times 2y^2x^2 \times 2xy =$

31) $4yx^4 \times 5y^5x \times xy^3 =$

32) $7x^2 \times 10x^3y^3 \times 8yx^4 =$

Multiplication Property of Exponents – Answers

✍ *Simplify and write the answer in exponential form.*

1) $2 \times 2^2 = 2^3$

2) $5^3 \times 5 = 5^4$

3) $3^2 \times 3^2 = 3^4$

4) $4^2 \times 4^2 = 4^4$

5) $7^3 \times 7^2 \times 7 = 7^6$

6) $2 \times 2^2 \times 2^2 = 2^5$

7) $5^3 \times 5^2 \times 5 \times 5 = 5^7$

8) $2x \times x = 2x^2$

9) $x^3 \times x^2 = x^5$

10) $x^4 \times x^4 = x^8$

11) $x^2 \times x^2 \times x^2 = x^6$

12) $6x \times 6x = 36x^2$

13) $2x^2 \times 2x^2 = 4x^4$

14) $3x^2 \times x = 3x^3$

15) $4x^4 \times 4x^4 \times 4x^4 = 64x^{12}$

16) $2x^2 \times x^2 = 2x^4$

17) $x^4 \times 3x = 3x^5$

18) $x \times 2x^2 = 2x^3$

19) $5x^4 \times 5x^4 = 25x^8$

20) $2yx^2 \times 2x = 4x^3y$

21) $3x^4 \times y^2x^4 = 3x^8y^2$

22) $y^2x^3 \times y^5x^2 = x^5y^7$

23) $4yx^3 \times 2x^2y^3 = 8x^5y^4$

24) $6x^2 \times 6x^3y^4 = 36x^5y^4$

25) $3x^4y^5 \times 7x^2y^3 = 21x^6y^8$

26) $7x^2y^5 \times 9xy^3 = 63x^3y^8$

27) $7xy^4 \times 4x^3y^3 = 28x^4y^7$

28) $3x^5y^3 \times 8x^2y^3 = 24x^7y^6$

29) $3x \times y^5x^3 \times y^4 = 3x^4y^9$

30) $yx^2 \times 2y^2x^2 \times 2xy = 4x^5y^4$

31) $4yx^4 \times 5y^5x \times xy^3 = 20x^6y^9$

32) $7x^2 \times 10x^3y^3 \times 8yx^4 = 560x^9y^4$

bit.ly/34AWHr1

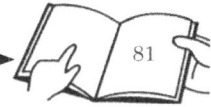

Division Property of Exponents

✎ *Simplify and write the answer in exponential form.*

1) $\dfrac{2^2}{2^3} =$

2) $\dfrac{2^4}{2^2} =$

3) $\dfrac{5^5}{5} =$

4) $\dfrac{3}{3^5} =$

5) $\dfrac{x}{x^3} =$

6) $\dfrac{3 \times 3^3}{3^2 \times 3^4} =$

7) $\dfrac{5^8}{5^3} =$

8) $\dfrac{5 \times 5^6}{5^2 \times 5^7} =$

9) $\dfrac{3^4 \times 3^7}{3^2 \times 3^8} =$

10) $\dfrac{5x}{10x^3} =$

11) $\dfrac{5x^3}{2x^5} =$

12) $\dfrac{18x^3}{14x^6} =$

13) $\dfrac{12x^3}{8xy^8} =$

14) $\dfrac{24xy^3}{4x^4y^2} =$

15) $\dfrac{21x^3y^9}{7xy^5} =$

16) $\dfrac{36x^2y^9}{4x^3} =$

17) $\dfrac{12x^4y^4}{10x^6y^7} =$

18) $\dfrac{12y^2x^{12}}{20yx^8} =$

19) $\dfrac{16x^4y}{9x^8y^2} =$

20) $\dfrac{5x^8y^2}{20x^5y^5} =$

Division Property of Exponents - Answers

✎ *Simplify and write the answer in exponential form.*

1) $\dfrac{2^2}{2^3} = \dfrac{1}{2}$

2) $\dfrac{2^4}{2^2} = 2^2$

3) $\dfrac{5^5}{5} = 5^4$

4) $\dfrac{3}{3^5} = \dfrac{1}{3^4}$

5) $\dfrac{x}{x^3} = \dfrac{1}{x^2}$

6) $\dfrac{3 \times 3^3}{3^2 \times 3^4} = \dfrac{1}{3}$

7) $\dfrac{5^8}{5^3} = 5^5$

8) $\dfrac{5 \times 5^6}{5^2 \times 5^7} = \dfrac{1}{5^2}$

9) $\dfrac{3^4 \times 3^7}{3^2 \times 3^8} = 3$

10) $\dfrac{5x}{10x^3} = \dfrac{1}{2x^2}$

11) $\dfrac{5x^3}{2x^5} = \dfrac{5}{2x^2}$

12) $\dfrac{18x^3}{14x^6} = \dfrac{9}{7x^3}$

13) $\dfrac{12x^3}{8xy^8} = \dfrac{3x^2}{2y^8}$

14) $\dfrac{24xy^3}{4x^4y^2} = \dfrac{6y}{x^3}$

15) $\dfrac{21x^3y^9}{7xy^5} = 3x^2y^4$

16) $\dfrac{36x^2y^9}{4x^3} = \dfrac{9y^9}{x}$

17) $\dfrac{12x^4y^4}{10x^6y^7} = \dfrac{6}{5x^2y^3}$

18) $\dfrac{12y^2x^{12}}{20yx^8} = \dfrac{3yx^4}{5}$

19) $\dfrac{16x^4y}{9x^8y^2} = \dfrac{16}{9x^4y}$

20) $\dfrac{5x^8y^2}{20x^5y^5} = \dfrac{x^3}{4y^3}$

bit.ly/37JAclZ

Powers of Products and Quotients

✍ *Simplify and write the answer in exponential form.*

1) $(4^2)^2 =$

2) $(6^2)^3 =$

3) $(2 \times 2^3)^4 =$

4) $(4 \times 4^4)^2 =$

5) $(3^3 \times 3^2)^3 =$

6) $(5^4 \times 5^5)^2 =$

7) $(2 \times 2^4)^2 =$

8) $(2x^6)^2 =$

9) $(11x^5)^2 =$

10) $(4x^2y^4)^4 =$

11) $(2x^4y^4)^3 =$

12) $(3x^2y^2)^2 =$

13) $(3x^4y^3)^4 =$

14) $(2x^6y^8)^2 =$

15) $(12x^3x)^3 =$

16) $(5x^9x^6)^3 =$

17) $(5x^{10}y^3)^3 =$

18) $(14x^3x^3)^2 =$

19) $(3x^3.5x)^2 =$

20) $(10x^{11}y^3)^2 =$

21) $(9x^7y^5)^2 =$

22) $(4x^4y^6)^5 =$

23) $(3x.4y^3)^2 =$

24) $(\frac{6x}{x^2})^2 =$

25) $\left(\frac{x^5y^5}{x^2y^2}\right)^3 =$

26) $\left(\frac{24x}{4x^6}\right)^2 =$

27) $\left(\frac{x^5}{x^7y^2}\right)^2 =$

28) $\left(\frac{xy^2}{x^2y^3}\right)^3 =$

29) $\left(\frac{4xy^4}{x^5}\right)^2 =$

30) $\left(\frac{xy^4}{5xy^2}\right)^3 =$

Powers of Products and Quotients - Answers

✍ *Simplify and write the answer in exponential form.*

1) $(4^2)^2 = 4^4$

2) $(6^2)^3 = 6^6$

3) $(2 \times 2^3)^4 = 2^{16}$

4) $(4 \times 4^4)^2 = 4^{10}$

5) $(3^3 \times 3^2)^3 = 3^{15}$

6) $(5^4 \times 5^5)^2 = 5^{18}$

7) $(2 \times 2^4)^2 = 2^{10}$

8) $(2x^6)^2 = 4x^{12}$

9) $(11x^5)^2 = 121x^{10}$

10) $(4x^2y^4)^4 = 256x^8y^{16}$

11) $(2x^4y^4)^3 = 8x^{12}y^{12}$

12) $(3x^2y^2)^2 = 9x^4y^4$

13) $(3x^4y^3)^4 = 81x^{16}y^{12}$

14) $(2x^6y^8)^2 = 4x^{12}y^{16}$

15) $(12x^3x)^3 = 1,728x^{12}$

16) $(5x^9x^6)^3 = 125x^{45}$

17) $(5x^{10}y^3)^3 = 125x^{30}y^9$

18) $(14x^3x^3)^2 = 196x^{12}$

19) $(3x^3.5x)^2 = 225x^8$

20) $(10x^{11}y^3)^2 = 100x^{22}y^6$

21) $(9x^7y^5)^2 = 81x^{14}y^{10}$

22) $(4x^4y^6)^5 = 1,024x^{20}y^{30}$

23) $(3x.4y^3)^2 = 144x^2y^6$

24) $\left(\frac{6x}{x^2}\right)^2 = \frac{36}{x^2}$

25) $\left(\frac{x^5y^5}{x^2y^2}\right)^3 = x^9y^9$

26) $\left(\frac{24x}{4x^6}\right)^2 = \frac{36}{x^{10}}$

27) $\left(\frac{x^5}{x^7y^2}\right)^2 = \frac{1}{x^4y^4}$

28) $\left(\frac{xy^2}{x^2y^3}\right)^3 = \frac{1}{x^3y^3}$

29) $\left(\frac{4xy^4}{x^5}\right)^2 = \frac{16y^8}{x^8}$

30) $\left(\frac{xy^4}{5xy^2}\right)^3 = \frac{y^6}{125}$

bit.ly/34CgPJm

Zero and Negative Exponents

 *Evaluate the following expressions.*

1) $1^{-1} =$

2) $2^{-2} =$

3) $0^{15} =$

4) $1^{-10} =$

5) $8^{-1} =$

6) $8^{-2} =$

7) $2^{-4} =$

8) $10^{-2} =$

9) $9^{-2} =$

10) $3^{-3} =$

11) $7^{-3} =$

12) $3^{-4} =$

13) $6^{-3} =$

14) $5^{-3} =$

15) $22^{-1} =$

16) $4^{-4} =$

17) $5^{-4} =$

18) $15^{-2} =$

19) $4^{-5} =$

20) $9^{-3} =$

21) $3^{-5} =$

22) $5^{-4} =$

23) $12^{-3} =$

24) $15^{-3} =$

25) $20^{-3} =$

26) $50^{-2} =$

27) $18^{-3} =$

28) $24^{-2} =$

29) $30^{-3} =$

30) $10^{-5} =$

31) $\left(\dfrac{1}{8}\right)^{-1}$

32) $\left(\dfrac{1}{5}\right)^{-2} =$

33) $\left(\dfrac{1}{7}\right)^{-2} =$

34) $\left(\dfrac{2}{3}\right)^{-2} =$

35) $\left(\dfrac{1}{5}\right)^{-3} =$

36) $\left(\dfrac{3}{4}\right)^{-2} =$

37) $\left(\dfrac{2}{5}\right)^{-2} =$

38) $\left(\dfrac{1}{2}\right)^{-8} =$

39) $\left(\dfrac{2}{5}\right)^{-3} =$

40) $\left(\dfrac{3}{7}\right)^{-2} =$

41) $\left(\dfrac{5}{6}\right)^{-3} =$

42) $\left(\dfrac{4}{9}\right)^{-2} =$

Zero and Negative Exponents - Answers

✍ *Evaluate the following expressions.*

1) $1^{-1} = 1$

2) $2^{-2} = \frac{1}{4}$

3) $0^{15} = 0$

4) $1^{-10} = 1$

5) $8^{-1} = \frac{1}{8}$

6) $8^{-2} = \frac{1}{64}$

7) $2^{-4} = \frac{1}{16}$

8) $10^{-2} = \frac{1}{100}$

9) $9^{-2} = \frac{1}{81}$

10) $3^{-3} = \frac{1}{27}$

11) $7^{-3} = \frac{1}{343}$

12) $3^{-4} = \frac{1}{81}$

13) $6^{-3} = \frac{1}{216}$

14) $5^{-3} = \frac{1}{125}$

15) $22^{-1} = \frac{1}{22}$

16) $4^{-4} = \frac{1}{256}$

17) $5^{-4} = \frac{1}{625}$

18) $15^{-2} = \frac{1}{225}$

19) $4^{-5} = \frac{1}{1,024}$

20) $9^{-3} = \frac{1}{729}$

21) $3^{-5} = \frac{1}{243}$

22) $5^{-4} = \frac{1}{625}$

23) $12^{-2} = \frac{1}{144}$

24) $15^{-3} = \frac{1}{3,375}$

25) $20^{-3} = \frac{1}{8,000}$

26) $50^{-2} = \frac{1}{2,500}$

27) $18^{-3} = \frac{1}{5,832}$

28) $24^{-2} = \frac{1}{576}$

29) $30^{-3} = \frac{1}{27,000}$

30) $10^{-5} = \frac{1}{100,000}$

31) $\left(\frac{1}{8}\right)^{-1} = 8$

32) $\left(\frac{1}{5}\right)^{-2} = 25$

33) $\left(\frac{1}{7}\right)^{-2} = 49$

34) $\left(\frac{2}{3}\right)^{-2} = \frac{9}{4}$

35) $\left(\frac{1}{5}\right)^{-3} = 125$

36) $\left(\frac{3}{4}\right)^{-2} = \frac{64}{27}$

37) $\left(\frac{2}{5}\right)^{-2} = \frac{25}{4}$

38) $\left(\frac{1}{2}\right)^{-8} = 256$

39) $\left(\frac{2}{5}\right)^{-3} = \frac{125}{8}$

40) $\left(\frac{3}{7}\right)^{-2} = \frac{49}{9}$

41) $\left(\frac{5}{6}\right)^{-3} = \frac{216}{125}$

42) $\left(\frac{4}{9}\right)^{-2} = \frac{81}{16}$

Find more at

bit.ly/3mkh4v

Negative Exponents and Negative Bases

✎ *Simplify and write the answer.*

1) $-3^{-1} =$

2) $-5^{-2} =$

3) $-2^{-4} =$

4) $-x^{-3} =$

5) $2x^{-1} =$

6) $-4x^{-3} =$

7) $-12x^{-5} =$

8) $-5x^{-2}y^{-3} =$

9) $20x^{-4}y^{-1} =$

10) $14a^{-6}b^{-7} =$

11) $-12x^2y^{-3} =$

12) $-\dfrac{25}{x^{-6}} =$

13) $-\dfrac{2x}{a^{-4}} =$

14) $\left(-\dfrac{1}{3x}\right)^{-2} =$

15) $\left(-\dfrac{3}{4x}\right)^{-2} =$

16) $-\dfrac{9}{a^{-7}b^{-2}} =$

17) $-\dfrac{5x}{x^{-3}} =$

18) $-\dfrac{a^{-3}}{b^{-2}} =$

19) $-\dfrac{8}{x^{-3}} =$

20) $\dfrac{5b}{-9c^{-4}} =$

21) $\dfrac{9ab}{a^{-3}b^{-1}} =$

22) $-\dfrac{15a^{-2}}{30b^{-3}} =$

23) $\dfrac{4ab^{-2}}{-3c^{-2}} =$

24) $\left(\dfrac{3a}{2c}\right)^{-2} =$

25) $\left(-\dfrac{5x}{3yz}\right)^{-3} =$

26) $\dfrac{11ab^{-2}}{-3c^{-2}} =$

27) $\left(-\dfrac{x^3}{x^4}\right)^{-2} =$

28) $\left(-\dfrac{x^{-2}}{3x^2}\right)^{-3} =$

Negative Exponents and Negative Bases - Answers

✎ Simplify and write the answer.

1) $-3^{-1} = -\frac{1}{3}$

2) $-5^{-2} = -\frac{1}{25}$

3) $-2^{-4} = -\frac{1}{16}$

4) $-x^{-3} = -\frac{1}{x^3}$

5) $2x^{-1} = \frac{2}{x}$

6) $-4x^{-3} = -\frac{4}{x^3}$

7) $-12x^{-5} = -\frac{12}{x^5}$

8) $-5x^{-2}y^{-3} = -\frac{5}{x^2y^3}$

9) $20x^{-4}y^{-1} = \frac{20}{x^4y}$

10) $14a^{-6}b^{-7} = \frac{14}{a^6b^7}$

11) $-12x^2y^{-3} = -\frac{12x^2}{y^3}$

12) $-\frac{25}{x^{-6}} = -25x^6$

13) $-\frac{2x}{a^{-4}} = -2xa^4$

14) $(-\frac{1}{3x})^{-2} = 9x^2$

15) $(-\frac{3}{4x})^{-2} = \frac{16x^2}{9}$

16) $-\frac{9}{a^{-7}b^{-2}} = -9a^7b^2$

17) $-\frac{5x}{x^{-3}} = -5x^4$

18) $-\frac{a^{-3}}{b^{-2}} = -\frac{b^2}{a^3}$

19) $-\frac{8}{x^{-3}} = -8x^3$

20) $\frac{5b}{-9c^{-4}} = -\frac{5bc^4}{9}$

21) $\frac{9ab}{a^{-3}b^{-1}} = 9a^4b^2$

22) $-\frac{15a^{-2}}{30b^{-3}} = -\frac{b^3}{2a^2}$

23) $\frac{4ab^{-2}}{-3c^{-2}} = -\frac{4ac^2}{3b^2}$

24) $(\frac{3a}{2c})^{-2} = \frac{4c^2}{9a^2}$

25) $(-\frac{5x}{3yz})^{-3} = -\frac{27y^3z^3}{125x^3}$

26) $\frac{11ab^{-2}}{-3c^{-2}} = -\frac{11ac^2}{3b^2}$

27) $(-\frac{x^3}{x^4})^{-2} = x^2$

28) $(-\frac{x^{-2}}{3x^2})^{-3} = -27x^{12}$

Scientific Notation

✍ *Write each number in scientific notation.*

1) $0.113 =$

2) $0.02 =$

3) $7.5 =$

4) $20 =$

5) $60 =$

6) $0.004 =$

7) $78 =$

8) $1,600 =$

9) $1,450 =$

10) $31,000 =$

11) $2,000,000 =$

12) $0.0000003 =$

13) $554,000 =$

14) $0.000725 =$

15) $0.00034 =$

16) $86,000,000 =$

17) $62,000 =$

18) $97,000,000 =$

19) $0.0000045 =$

20) $0.0019 =$

✍ *Write each number in standard notation.*

1) $2 \times 10^{-1} =$

2) $8 \times 10^{-2} =$

3) $1.8 \times 10^3 =$

4) $9 \times 10^{-4} =$

5) $1.7 \times 10^{-2} =$

6) $9 \times 10^3 =$

7) $7 \times 10^5 =$

8) $1.15 \times 10^4 =$

9) $7 \times 10^{-5} =$

10) $8.3 \times 10^{-5} =$

Scientific Notation - Answers

✎ *Write each number in scientific notation.*

1) $0.113 = 1.13 \times 10^{-1}$

2) $0.02 = 2 \times 10^{-2}$

3) $7.5 = 2.5 \times 10^{0}$

4) $20 = 2 \times 10^{1}$

5) $60 = 6 \times 10^{1}$

6) $0.004 = 4 \times 10^{-3}$

7) $78 = 7.8 \times 10^{1}$

8) $1,600 = 1.6 \times 10^{3}$

9) $1,450 = 1.45 \times 10^{3}$

10) $31,000 = 3.1 \times 10^{4}$

11) $2,000,000 = 2 \times 10^{6}$

12) $0.0000003 = 3 \times 10^{-7}$

13) $554,000 = 5.54 \times 10^{5}$

14) $0.000725 = 7.25 \times 10^{-4}$

15) $0.00034 = 3.4 \times 10^{-4}$

16) $86,000,000 = 8.6 \times 10^{7}$

17) $62,000 = 6.2 \times 10^{4}$

18) $97,000,000 = 9.7 \times 10^{7}$

19) $0.0000045 = 4.5 \times 10^{-6}$

20) $0.0019 = 1.9 \times 10^{-3}$

✎ *Write each number in standard notation.*

1) $2 \times 10^{-1} = 0.2$

2) $8 \times 10^{-2} = 0.08$

3) $1.8 \times 10^{3} = 1,800$

4) $9 \times 10^{-4} = 0.0009$

5) $1.7 \times 10^{-2} = 0.017$

6) $9 \times 10^{3} = 9,000$

7) $7 \times 10^{5} = 700,000$

8) $1.15 \times 10^{4} = 11,500$

9) $7 \times 10^{-5} = 0.00007$

10) $8.3 \times 10^{-5} = 0.000083$

Radicals

Simplify and write the answer.

1) $\sqrt{0} =$ ____

2) $\sqrt{1} =$ ____

3) $\sqrt{4} =$ ____

4) $\sqrt{16} =$ ____

5) $\sqrt{9} =$ ____

6) $\sqrt{25} =$ ____

7) $\sqrt{49} =$ ____

8) $\sqrt{36} =$ ____

9) $\sqrt{64} =$ ____

10) $\sqrt{81} =$ ____

11) $\sqrt{121} =$ ____

12) $\sqrt{225} =$ ____

13) $\sqrt{144} =$ ____

14) $\sqrt{100} =$ ____

15) $\sqrt{256} =$ ____

16) $\sqrt{289} =$ ____

17) $\sqrt{324} =$ ____

18) $\sqrt{400} =$ ____

19) $\sqrt{900} =$ ____

20) $\sqrt{529} =$ ____

21) $\sqrt{361} =$ ____

22) $\sqrt{169} =$ ____

23) $\sqrt{196} =$ ____

24) $\sqrt{90} =$ ____

Evaluate.

1) $\sqrt{6} \times \sqrt{6} =$

2) $\sqrt{5} \times \sqrt{5} =$

3) $\sqrt{8} \times \sqrt{8} =$

4) $\sqrt{2} + \sqrt{2} =$

5) $\sqrt{8} + \sqrt{8} =$

6) $6\sqrt{5} - 2\sqrt{5} =$

7) $\sqrt{25} \times \sqrt{16} =$

8) $\sqrt{25} \times \sqrt{64} =$

9) $\sqrt{81} \times \sqrt{25} =$

10) $5\sqrt{3} \times 2\sqrt{3} =$

11) $8\sqrt{2} \times 2\sqrt{2} =$

12) $6\sqrt{3} - \sqrt{12} =$

Radicals - Answers

✍ *Simplify and write the answer.*

1) $\sqrt{0} = 0$

2) $\sqrt{1} = 1$

3) $\sqrt{4} = 2$

4) $\sqrt{16} = 4$

5) $\sqrt{9} = 3$

6) $\sqrt{25} = 5$

7) $\sqrt{49} = 7$

8) $\sqrt{36} = 6$

9) $\sqrt{64} = 8$

10) $\sqrt{81} = 9$

11) $\sqrt{121} = 11$

12) $\sqrt{225} = 15$

13) $\sqrt{144} = 12$

14) $\sqrt{100} = 10$

15) $\sqrt{256} = 16$

16) $\sqrt{289} = 17$

17) $\sqrt{324} = 18$

18) $\sqrt{400} = 20$

19) $\sqrt{900} = 30$

20) $\sqrt{529} = 23$

21) $\sqrt{361} = 19$

22) $\sqrt{169} = 13$

23) $\sqrt{196} = 14$

24) $\sqrt{90} = 3\sqrt{10}$

✍ *Evaluate.*

1) $\sqrt{6} \times \sqrt{6} = 6$

2) $\sqrt{5} \times \sqrt{5} = 5$

3) $\sqrt{8} \times \sqrt{8} = 8$

4) $\sqrt{2} + \sqrt{2} = 2\sqrt{2}$

5) $\sqrt{8} + \sqrt{8} = 2\sqrt{8} = 4\sqrt{2}$

6) $6\sqrt{5} - 2\sqrt{5} = 4\sqrt{5}$

7) $\sqrt{25} \times \sqrt{16} = 20$

8) $\sqrt{25} \times \sqrt{64} = 40$

9) $\sqrt{81} \times \sqrt{25} = 45$

10) $5\sqrt{3} \times 2\sqrt{3} = 30$

11) $8\sqrt{2} \times 2\sqrt{2} = 32$

12) $6\sqrt{3} - \sqrt{12} = 4\sqrt{3}$

Simplifying Polynomials

✏️ *Simplify each expression.*

1) $2(2x + 2) =$

_____

2) $4(4x - 2) =$

_____

3) $3(5x + 3) =$

_____

4) $6(7x + 5) =$

_____

5) $-3(8x - 7) =$

_____

6) $2x(3x + 4) =$

_____

7) $3x^2 + 3x^2 - 2x^3 =$

_____

8) $2x - x^2 + 6x^3 + 4 =$

_____

9) $5x + 2x^2 - 9x^3 =$

_____

10) $7x^2 + 5x^4 - 2x^3 =$

_____

11) $-3x^2 + 5x^3 + 6x^4 =$

_____

12) $(x - 3)(x - 4) =$

_____

13) $(x - 5)(x + 4) =$

_____

14) $(x - 6)(x - 3) =$

_____

15) $(2x + 5)(x + 8) =$

_____

16) $(3x - 8)(x + 4) =$

_____

17) $-8x^2 + 2x^3 - 10x^4 + 5x =$

_____

18) $11 - 6x^2 + 5x^2 - 12x^3 + 22 =$

_____

19) $2x^2 - 2x + 3x^3 + 12x - 22x =$

_____

20) $11 - 4x^2 + 3x^2 - 7x^3 + 3 =$

_____

21) $2x^5 - x^3 + 8x^2 - 2x^5 =$

_____

22) $(2x^3 - 1) + (3x^3 - 2x^3) =$

_____

Simplifying Polynomials - Answers

✎ *Simplify each expression.*

1) $2(2x + 2) =$
   $4x + 4$

2) $4(4x - 2) =$
   $16x - 8$

3) $3(5x + 3) =$
   $15x + 9$

4) $6(7x + 5) =$
   $42x + 30$

5) $-3(8x - 7) =$
   $-24x + 21$

6) $2x(3x + 4) =$
   $6x^2 + 8x$

7) $3x^2 + 3x^2 - 2x^3 =$
   $-2x^3 + 6x^2$

8) $2x - x^2 + 6x^3 + 4 =$
   $6x^3 - x^2 + 2x + 4$

9) $5x + 2x^2 - 9x^3 =$
   $-9x^3 + 2x^2 + 5x$

10) $7x^2 + 5x^4 - 2x^3 =$
    $5x^4 - 2x^3 + 7x^2$

11) $-3x^2 + 5x^3 + 6x^4 =$
    $6x^4 + 5x^3 - 3x^2$

12) $(x - 3)(x - 4) =$
    $x^2 - 7x + 12$

13) $(x - 5)(x + 4) =$
    $x^2 - x - 20$

14) $(x - 6)(x - 3) =$
    $x^2 - 9x + 18$

15) $(2x + 5)(x + 8) =$
    $2x^2 + 21x + 40$

16) $(3x - 8)(x + 4) =$
    $3x^2 + 4x - 32$

17) $-8x^2 + 2x^3 - 10x^4 + 5x =$
    $-10x^4 + 2x^3 - 8x^2 + 5x$

18) $11 - 6x^2 + 5x^2 - 12x^3 + 22 =$
    $-12x^3 - x^2 + 33$

19) $2x^2 - 2x + 3x^3 + 12x - 22x =$
    $3x^3 + 2x^2 - 12x$

20) $11 - 4x^2 + 3x^2 - 7x^3 + 3 =$
    $-7x^3 - x^2 + 14$

21) $2x^5 - x^3 + 8x^2 - 2x^5 =$
    $-x^3 + 8x^2$

22) $(2x^3 - 1) + (3x^3 - 2x^3) =$
    $3x^3 - 1$

Adding and Subtracting Polynomials

✍ *Add or subtract expressions.*

1) $(x^2 - 3) + (x^2 + 1) =$

_____

2) $(2x^2 - 4) - (2 - 4x^2) =$

_____

3) $(x^3 + 2x^2) - (x^3 + 5) =$

_____

4) $(3x^3 - x^2) + (4x^2 - 7x) =$

_____

5) $(2x^3 + 3x) - (5x^3 + 2) =$

_____

6) $(5x^3 - 2) + (2x^3 + 10) =$

_____

7) $(7x^3 + 5) - (9 - 4x^3) =$

_____

8) $(5x^2 + 3x^3) - (2x^3 + 6) =$

_____

9) $(8x^2 - x) + (4x - 8x^2) =$

_____

10) $(6x + 9x^2) - (5x + 2) =$

_____

11) $(7x^4 - 2x) - (6x - 2x^4) =$

_____

12) $(2x - 4x^3) - (9x^3 + 6x) =$

_____

13) $(8x^3 - 8x^2) - (6x^2 - 3x) =$

_____

14) $(9x^2 - 6) + (5x^2 - 4x^3) =$

_____

15) $(8x^3 + 3x^4) - (x^4 - 3x^3) =$

_____

16) $(-4x^3 - 2x) + (5x - 2x^3) =$

_____

17) $(6x - 4x^4) - (8x^4 + 3x) =$

_____

18) $(7x - 8x^2) - (9x^4 - 3x^2) =$

_____

19) $(9x^3 - 6) + (9x^3 - 5x^2) =$

_____

20) $(5x^3 + x^4) - (8x^4 - 7x^3) =$

_____

Adding and Subtracting Polynomials - Answers

✎ *Add or subtract expressions.*

1) $(x^2 - 3) + (x^2 + 1) =$

$2x^2 - 2$

2) $(2x^2 - 4) - (2 - 4x^2) =$

$6x^2 - 6$

3) $(x^3 + 2x^2) - (x^3 + 5) =$

$2x^2 - 5$

4) $(3x^3 - x^2) + (4x^2 - 7x) =$

$3x^3 + 3x^2 - 7x$

5) $(2x^3 + 3x) - (5x^3 + 2) =$

$-3x^3 + 3x^2 - 2$

6) $(5x^3 - 2) + (2x^3 + 10) =$

$7x^3 + 8$

7) $(7x^3 + 5) - (9 - 4x^3) =$

$11x^3 - 4$

8) $(5x^2 + 3x^3) - (2x^3 + 6) =$

$x^3 + 5x^2 - 6$

9) $(8x^2 - x) + (4x - 8x^2) =$

$3x$

10) $(6x + 9x^2) - (5x + 2) =$

$9x^2 + x - 2$

11) $(7x^4 - 2x) - (6x - 2x^4) =$

$9x^4 - 8x$

12) $(2x - 4x^3) - (9x^3 + 6x) =$

$-13x^3 - 4x$

13) $(8x^3 - 8x^2) - (6x^2 - 3x) =$

$8x^3 - 14x^2 + 3x$

14) $(9x^2 - 6) + (5x^2 - 4x^3) =$

$-4x^3 + 14x^2 - 6$

15) $(8x^3 + 3x^4) - (x^4 - 3x^3) =$

$2x^4 + 11x^3$

16) $(-4x^3 - 2x) + (5x - 2x^3) =$

$-6x^3 + 3x$

17) $(6x - 4x^4) - (8x^4 + 3x) =$

$-12x^4 + 3x$

18) $(7x - 8x^2) - (9x^4 - 3x^2) =$

$-9x^4 - 5x^2 + 7x$

19) $(9x^3 - 6) + (9x^3 - 5x^2) =$

$18x^3 - 5x^2 - 6$

20) $(5x^3 + x^4) - (8x^4 - 7x^3) =$

$-7x^4 + 12x^3$

Multiplying Monomials

✍ *Simplify each expression.*

1) $5x^8 \times x^3 =$

2) $-4z^7 \times 5z^5 =$

3) $-6xy^8 \times 3x^5y^3 =$

4) $5xy^5 \times 3x^3y^4 =$

5) $8s^6t^2 \times 6s^3t^7 =$

6) $9xy^6z \times 3y^4z^2 =$

7) $4pq^5 \times (-7p^4q^8) =$

8) $10p^3q^5 \times (-4p^4q^6) =$

9) $(-9a^4b^7c^4) \times (-4a^7b) =$

10) $5u^3v^9z^2 \times (-4uv^9z) =$

11) $8x^3y^2z^5 \times (-9x^4y^2z) =$

12) $5y^5 \times 6y^3 =$

13) $7x^5y \times 3xy^2 =$

14) $7a^4b^2 \times 3a^8b =$

15) $5p^5q^4 \times (-6pq^4) =$

16) $(-8x^5y^2) \times 4x^6y^3 =$

17) $12x^5y^4 \times 2x^8y =$

18) $9s^4t^2 \times (-5st^5) =$

19) $(-5p^2q^4r) \times 7pq^5r^3 =$

20) $7u^5v^9 \times (-5u^{12}v^7) =$

21) $(-9xy^2z^4) \times 2x^2yz^5 =$

22) $6a^8b^8c^{12} \times 9a^7b^5c^8 =$

Multiplying Monomials - Answers

✍ *Simplify each expression.*

1) $5x^8 \times x^3 =$

$5x^{11}$

2) $-4z^7 \times 5z^5 =$

$-20z^{12}$

3) $-6xy^8 \times 3x^5y^3 =$

$-18x^6y^{11}$

4) $5xy^5 \times 3x^3y^4 =$

$15x^4y^9$

5) $8s^6t^2 \times 6s^3t^7 =$

$48s^9t^9$

6) $9xy^6z \times 3y^4z^2 =$

$27xy^{10}z^3$

7) $4pq^5 \times (-7p^4q^8) =$

$-28p^5q^{13}$

8) $10p^3q^5 \times (-4p^4q^6) =$

$-40p^7q^{11}$

9) $(-9a^4b^7c^4) \times (-4a^7b) =$

$36a^{11}b^8c^4$

10) $5u^3v^9z^2 \times (-4uv^9z) =$

$-20u^4v^{18}z^3$

11) $8x^3y^2z^5 \times (-9x^4y^2z) =$

$-72x^7y^4z^6$

12) $5y^5 \times 6y^3 =$

$30y^8$

13) $7x^5y \times 3xy^2 =$

$21x^6y^3$

14) $7a^4b^2 \times 3a^8b =$

$21a^{12}b^3$

15) $5p^5q^4 \times (-6pq^4) =$

$-30p^6q^8$

16) $(-8x^5y^2) \times 4x^6y^3 =$

$-32x^{11}y^5$

17) $12x^5y^4 \times 2x^8y =$

$24x^{13}y^5$

18) $9s^4t^2 \times (-5st^5) =$

$-45s^5t^7$

19) $(-5p^2q^4r) \times 7pq^5r^3 =$

$-35p^3q^9r^4$

20) $7u^5v^9 \times (-5u^{12}v^7) =$

$-35u^{17}v^{16}$

21) $(-9xy^2z^4) \times 2x^2yz^5 =$

$-18x^3y^3z^9$

22) $6a^8b^8c^{12} \times 9a^7b^5c^8 =$

$54a^{15}b^{13}c^{20}$

Multiplying and Dividing Monomials

✍️ *Simplify each expression.*

1) $(8x^3)(2x^2) =$

2) $(4x^6)(5x^4) =$

3) $(-6x^8)(3x^3) =$

4) $(5x^8y^9)(-6x^6y^9) =$

5) $(8x^5y^6)(3x^2y^5) =$

6) $(8yx^2)(7y^5x^3) =$

7) $(4x^2y)(2x^2y^3) =$

8) $(-2x^9y^4)(-9x^6y^8) =$

9) $(-5x^8y^2)(-6x^4y^5) =$

10) $(8x^8y)(-7x^4y^3) =$

11) $(9x^6y^2)(6x^7y^4) =$

12) $(8x^9y^5)(6x^5y^4) =$

13) $(-5x^8y^9)(7x^7y^8) =$

14) $(6x^2y^5)(5x^3y^2) =$

15) $(9x^5y^{12})(4x^7y^9) =$

16) $(-10x^{14}y^8)(2x^7y^5) =$

17) $\dfrac{8x^4y^3}{xy^2} =$

18) $\dfrac{6x^5y^6}{2x^3y} =$

19) $\dfrac{12x^3y^7}{4xy} =$

20) $\dfrac{-20x^8y^9}{5x^5y^4} =$

Multiplying and Dividing Monomials - Answers

✏️ *Simplify each expression.*

1) $(8x^3)(2x^2) =$

$16x^5$

2) $(4x^6)(5x^4) =$

$20x^{10}$

3) $(-6x^8)(3x^3) =$

$-18x^{11}$

4) $(5x^8y^9)(-6x^6y^9) =$

$-30x^{14}y^{18}$

5) $(8x^5y^6)(3x^2y^5) =$

$24x^7y^{11}$

6) $(8yx^2)(7y^5x^3) =$

$56y^6x^5$

7) $(4x^2y)(2x^2y^3) =$

$8x^4y^4$

8) $(-2x^9y^4)(-9x^6y^8) =$

$18x^{15}y^{12}$

9) $(-5x^8y^2)(-6x^4y^5) =$

$30x^{12}y^7$

10) $(8x^8y)(-7x^4y^3) =$

$-56x^{12}y^4$

11) $(9x^6y^2)(6x^7y^4) =$

$54x^{13}y^6$

12) $(8x^9y^5)(6x^5y^4) =$

$48x^{14}y^9$

13) $(-5x^8y^9)(7x^7y^8) =$

$-35x^{15}y^{17}$

14) $(6x^2y^5)(5x^3y^2) =$

$30x^5y^7$

15) $(9x^5y^{12})(4x^7y^9) =$

$36x^{12}y^{21}$

16) $(-10x^{14}y^8)(2x^7y^5) =$

$-20x^{21}y^{13}$

17) $\dfrac{8x^4y^3}{xy^2} =$

$8x^3y$

18) $\dfrac{6x^5y^6}{2x^3y} =$

$3x^2y^5$

19) $\dfrac{12x^3y^7}{4xy} =$

$3x^2y^6$

20) $\dfrac{-20x^8y^9}{5x^5y^4} =$

$-4x^3y^5$

Multiplying a Polynomial and a Monomial

✎ *Find each product.*

1) $x(x - 2) =$

   _____

2) $2(2 + x) =$

   _____

3) $x(x - 1) =$

   _____

4) $x(x + 3) =$

   _____

5) $2x(x - 2) =$

   _____

6) $5(4x + 3) =$

   _____

7) $4x(3x - 4) =$

   _____

8) $x(5x + 2y) =$

   _____

9) $3x(x - 2y) =$

   _____

10) $6x(3x - 4y) =$

   _____

11) $2x(3x - 8) =$

   _____

12) $6x(4x - 6y) =$

   _____

13) $3x(4x - 2y) =$

   _____

14) $2x(2x - 6y) =$

   _____

15) $5x(x^2 + y^2) =$

   _____

16) $3x(2x^2 - y^2) =$

   _____

17) $7(2x^2 + 9y^2) =$

   _____

18) $2x(-2x^2y + 3y) =$

   _____

19) $-2(2x^2 - 4xy + 2) =$

   _____

20) $5(x^2 - 6xy - 8) =$

   _____

bit.ly/3aBYdx2

Find more at

Multiplying a Polynomial and a Monomial - Answers

✍ *Find each product.*

1) $x(x - 2) =$

$x^2 - 2x$

2) $2(2 + x) =$

$2x + 4$

3) $x(x - 1) =$

$x^2 - x$

4) $x(x + 3) =$

$x^2 + 3x$

5) $2x(x - 2) =$

$2x^2 - 4x$

6) $5(4x + 3) =$

$20x + 15$

7) $4x(3x - 4) =$

$12x^2 - 16x$

8) $x(5x + 2y) =$

$5x^2 + 2xy$

9) $3x(x - 2y) =$

$3x^2 - 6xy$

10) $6x(3x - 4y) =$

$18x^2 - 24xy$

11) $2x(3x - 8) =$

$6x^2 - 16x$

12) $6x(4x - 6y) =$

$24x^2 - 36xy$

13) $3x(4x - 2y) =$

$12x^2 - 6xy$

14) $2x(2x - 6y) =$

$4x^2 - 12xy$

15) $5x(x^2 + y^2) =$

$5x^3 + 5xy^2$

16) $3x(2x^2 - y^2) =$

$6x^3 - 3xy^2$

17) $7(2x^2 + 9y^2) =$

$14x^2 + 63y^2$

18) $2x(-2x^2y + 3y) =$

$-4x^3y + 6xy$

19) $-2(2x^2 - 4xy + 2) =$

$-4x^2 + 8xy - 4$

20) $5(x^2 - 6xy - 8) =$

$5x^2 - 30xy - 40$

Multiplying Binomials

✍ *Find each product.*

1) $(x - 2)(x + 5) =$

_____

2) $(x + 4)(x + 2) =$

_____

3) $(x - 2)(x - 4) =$

_____

4) $(x - 8)(x - 2) =$

_____

5) $(x - 7)(x - 5) =$

_____

6) $(x + 6)(x + 2) =$

_____

7) $(x - 9)(x + 3) =$

_____

8) $(x - 8)(x - 5) =$

_____

9) $(x + 3)(x + 7) =$

_____

10) $(x - 9)(x + 4) =$

_____

11) $(x + 6)(x + 6) =$

_____

12) $(x + 7)(x + 7) =$

_____

13) $(x - 8)(x + 7) =$

_____

14) $(x + 9)(x + 9) =$

_____

15) $(x - 8)(x - 8) =$

_____

16) $(2x - 9)(x + 5) =$

_____

17) $(2x - 3)(x + 4) =$

_____

18) $(2x + 4)(x + 2) =$

_____

19) $(2x + 2)(x + 3) =$

_____

20) $(2x - 4)(2x + 2) =$

_____

bit.ly/3aCsOFL

Find more at

Multiplying Binomials - Answers

✍️ *Find each product.*

1) $(x - 2)(x + 5) =$

   $x^2 + 3x - 10$

2) $(x + 4)(x + 2) =$

   $x^2 + 6x + 8$

3) $(x - 2)(x - 4) =$

   $x^2 - 6x + 8$

4) $(x - 8)(x - 2) =$

   $x^2 - 10x + 16$

5) $(x - 7)(x - 5) =$

   $x^2 - 12x + 35$

6) $(x + 6)(x + 2) =$

   $x^2 + 8x + 12$

7) $(x - 9)(x + 3) =$

   $x^2 - 6x - 27$

8) $(x - 8)(x - 5) =$

   $x^2 - 13x + 40$

9) $(x + 3)(x + 7) =$

   $x^2 + 10x + 21$

10) $(x - 9)(x + 4) =$

   $x^2 - 5x - 36$

11) $(x + 6)(x + 6) =$

   $x^2 + 12x + 36$

12) $(x + 7)(x + 7) =$

   $x^2 + 14x + 49$

13) $(x - 8)(x + 7) =$

   $x^2 - x - 56$

14) $(x + 9)(x + 9) =$

   $x^2 + 18x + 81$

15) $(x - 8)(x - 8) =$

   $x^2 - 16x + 64$

16) $(2x - 9)(x + 5) =$

   $2x^2 + x - 45$

17) $(2x - 3)(x + 4) =$

   $2x^2 + 5x - 12$

18) $(2x + 4)(x + 2) =$

   $2x^2 + 8x + 8$

19) $(2x + 2)(x + 3) =$

   $2x^2 + 8x + 6$

20) $(2x - 4)(2x + 2) =$

   $4x^2 - 4x - 8$

Factoring Trinomials

✍ *Factor each trinomial.*

1) $x^2 + 3x - 10 =$

_____

2) $x^2 + 6x + 8 =$

_____

3) $x^2 - 6x + 8 =$

_____

4) $x^2 - 10x + 16 =$

_____

5) $x^2 - 13x + 40 =$

_____

6) $x^2 + 8x + 12 =$

_____

7) $x^2 - 6x - 27 =$

_____

8) $x^2 - 14x + 48 =$

_____

9) $x^2 + 15x + 56 =$

_____

10) $x^2 - 5x - 36 =$

_____

11) $x^2 + 12x + 36 =$

_____

12) $x^2 + 16x + 63 =$

_____

13) $x^2 + x - 72 =$

_____

14) $x^2 + 18x + 81 =$

_____

15) $x^2 - 16x + 64 =$

_____

16) $x^2 - 18x + 81 =$

_____

17) $2x^2 + 8x + 6 =$

_____

18) $2x^2 + 6x - 8 =$

_____

19) $2x^2 + 12x + 10 =$

_____

20) $4x^2 + 6x - 28 =$

_____

Factoring Trinomials – Answer

✎ *Factor each trinomial.*

1) $x^2 + 3x - 10 =$

$(x - 2)(x + 5)$

2) $x^2 + 6x + 8 =$

$(x + 4)(x + 2)$

3) $x^2 - 6x + 8 =$

$(x - 2)(x - 4)$

4) $x^2 - 10x + 16 =$

$(x - 8)(x - 2)$

5) $x^2 - 13x + 40 =$

$(x - 8)(x - 5)$

6) $x^2 + 8x + 12 =$

$(x + 6)(x + 2)$

7) $x^2 - 6x - 27 =$

$(x - 9)(x + 3)$

8) $x^2 - 14x + 48 =$

$(x - 8)(x - 6)$

9) $x^2 + 15x + 56 =$

$(x + 8)(x + 7)$

10) $x^2 - 5x - 36 =$

$(x - 9)(x + 4)$

11) $x^2 + 12x + 36 =$

$(x + 6)(x + 6)$

12) $x^2 + 16x + 63 =$

$(x + 7)(x + 9)$

13) $x^2 + x - 72 =$

$(x - 8)(x + 9)$

14) $x^2 + 18x + 81 =$

$(x + 9)(x + 9)$

15) $x^2 - 16x + 64 =$

$(x - 8)(x - 8)$

16) $x^2 - 18x + 81 =$

$(x - 9)(x - 9)$

17) $2x^2 + 8x + 6 =$

$(2x + 2)(x + 3)$

18) $2x^2 + 6x - 8 =$

$(2x - 2)(x + 4)$

19) $2x^2 + 12x + 10 =$

$(2x + 2)(x + 5)$

20) $4x^2 + 6x - 28 =$

$(2x - 4)(2x + 7)$

The Pythagorean Theorem

✍ *Do the following lengths form a right triangle?*

1) _____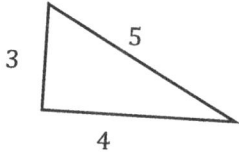
3, 5, 4

2) _____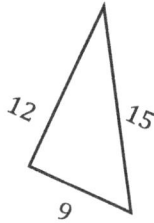
12, 15, 9

3) _____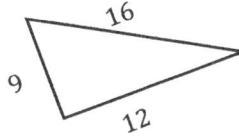
16, 9, 12

4) _____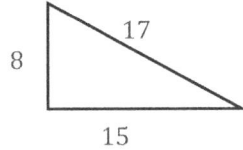
17, 8, 15

5) _____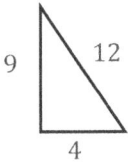
9, 12, 4

6) _____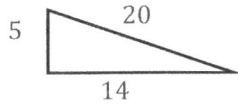
20, 5, 14

7) _____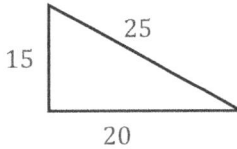
25, 15, 20

8) _____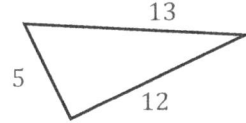
13, 5, 12

✍ *Find the missing side.*

9) _____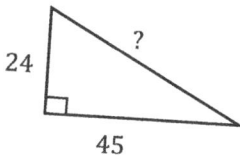
24, ?, 45

10) _____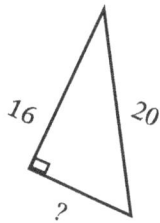
16, 20, ?

11) _____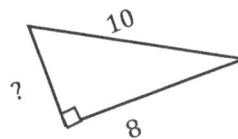
10, ?, 8

12) _____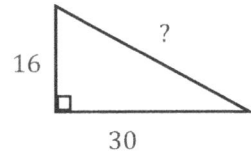
16, ?, 30

13) _____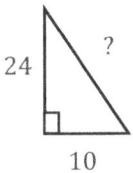
24, ?, 10

14) _____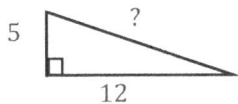
5, ?, 12

15) _____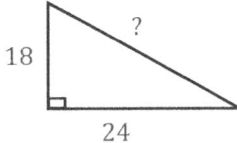
18, ?, 24

16) _____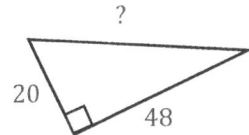
?, 20, 48

bit.ly/37H08y

The Pythagorean Theorem - Answers

✎ *Do the following lengths form a right triangle?*

1) yes

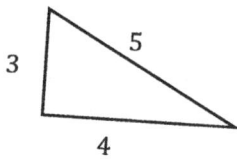

5
3
4

2) yes

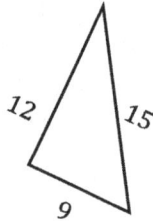

12   15
9

3) no

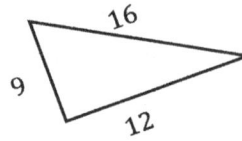

16
9
12

4) yes

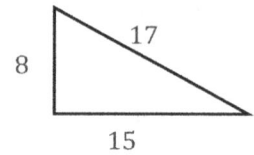

17
8
15

5) no

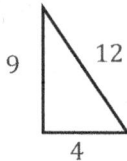

9   12
4

6) no

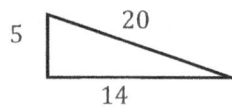

5   20
14

7) yes

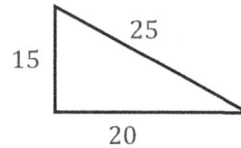

25
15
20

8) yes

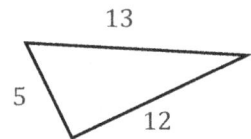

13
5
12

✎ *Find the missing side.*

9) 51

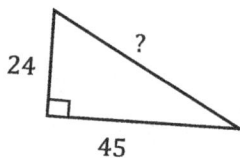

24   ?
45

10) 12

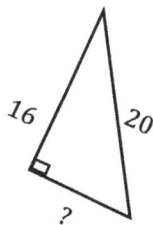

16   20
?

11) 6

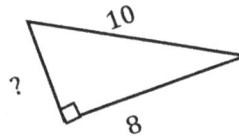

10
?
8

12) 34

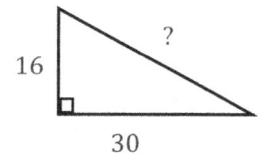

?
16
30

13) 26

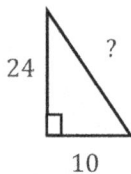

24   ?
10

14) 13

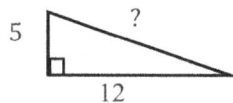

?
5
12

15) 30

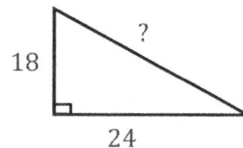

?
18
24

16) 52

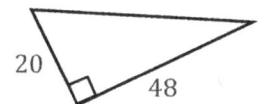

?
20
48

Triangles

✎ **Find the measure of the unknown angle in each triangle.**

1) _____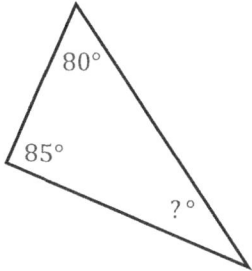
80° 85° ?°

2) _____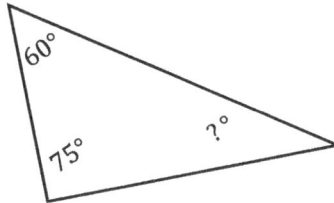
60° 75° ?°

3) _____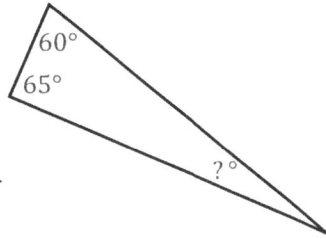
60° 65° ?°

4) _____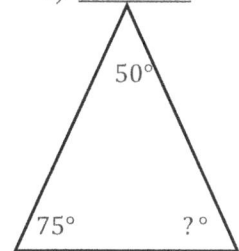
50° 75° ?°

5) _____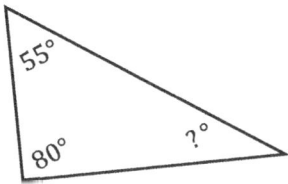
55° 80° ?°

6) _____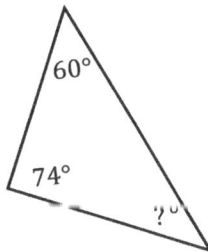
60° 74° ?°

7) _____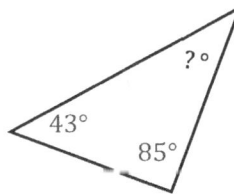
?° 43° 85°

8) _____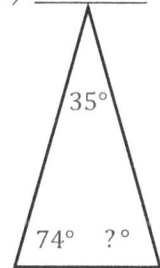
35° 74° ?°

✎ **Find area of each triangle.**

9) _____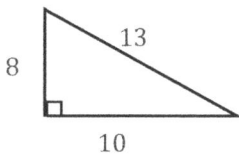
13 8 10

10) _____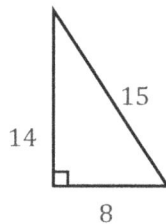
15 14 8

11) _____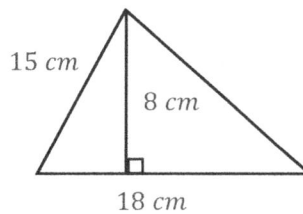
15 cm 8 cm 18 cm

12) _____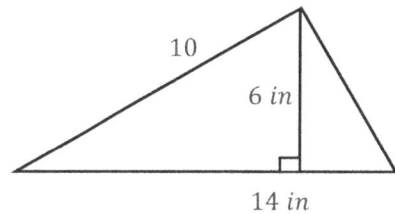
10 6 in 14 in

EffortlessMath.com

 Find more at bit.ly/3haZrRg

Triangles – Answers

✎ *Find the measure of the unknown angle in each triangle.*

1) 15°

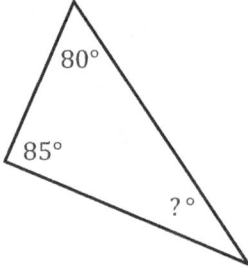

80°

85°

?°

2) 45°

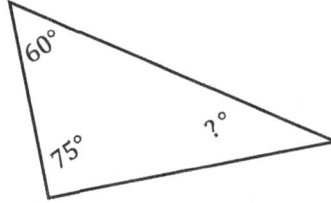

60°

75°

?°

3) 55°

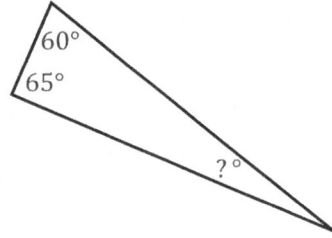

60°

65°

?°

4) 55°

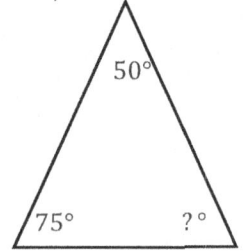

50°

75°

?°

5) 45°

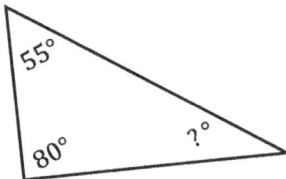

55°

80°

?°

6) 46°

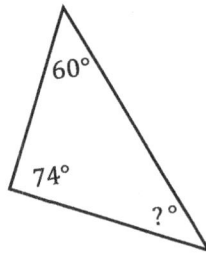

60°

74°

?°

7) 52°

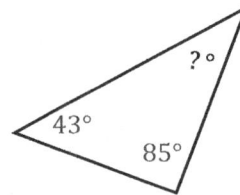

?°

43°

85°

8) 71°

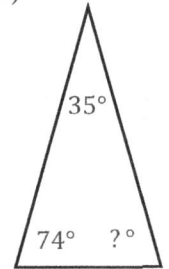

35°

74° ?°

✎ *Find area of each triangle.*

9) 40

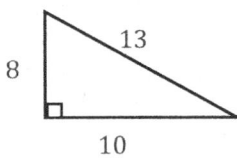

8

13

10

10) 56

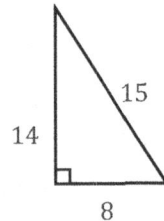

14

15

8

11) 72 cm^2

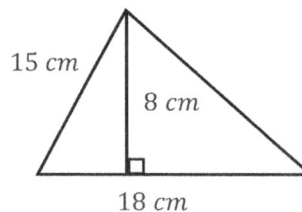

15 *cm*

8 *cm*

18 *cm*

12) 42 in^2

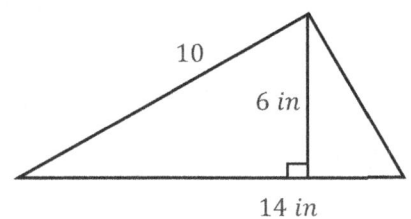

10

6 *in*

14 *in*

Find more at

bit.ly/3haZrRg

EffortlessMath.com

Polygons

✍️ *Find the perimeter of each shape.*

1) (square) _____

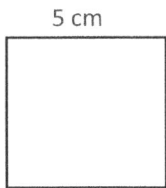

5 cm

2) _____

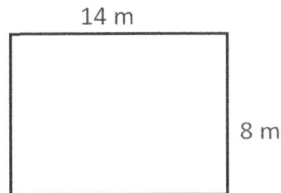

14 m

8 m

3) _____

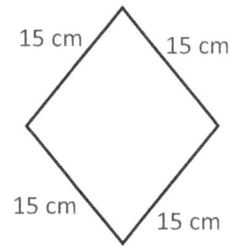

15 cm   15 cm

15 cm   15 cm

4) (square) _____

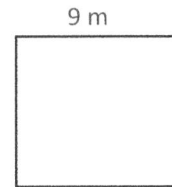

9 m

5) *(regular hexagon)* _____

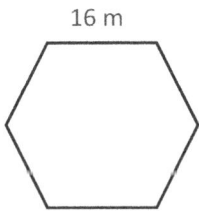

16 m

6) _____

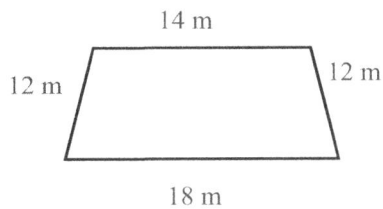

14 m

12 m   12 m

18 m

7) *(parallelogram)* _____

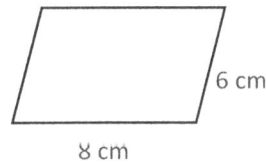

6 cm

8 cm

8) *(regular hexagon)* _____

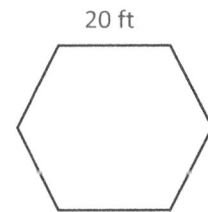

20 ft

9) _____

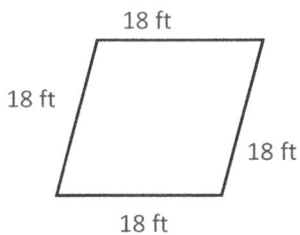

18 ft

18 ft

18 ft

18 ft

10) _____

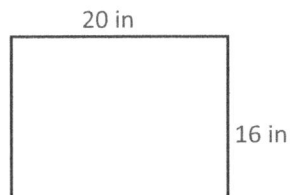

20 in

16 in

11) _____

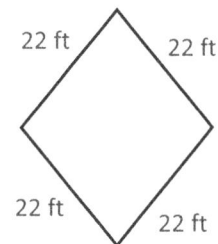

22 ft   22 ft

22 ft   22 ft

12) *(regular hexagon)* _____

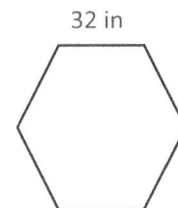

32 in

bit.ly/3nFNiGi

Find more at

Polygons - Answers

✎ *Find the perimeter of each shape.*

1) (square) *20 cm*   2) *44 m*   3) *60 cm*   4) (square) *36 m*

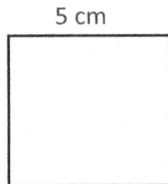
5 cm

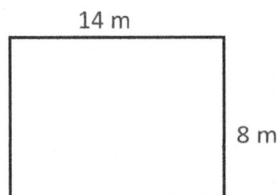
14 m
8 m

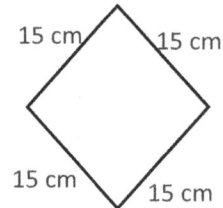
15 cm 15 cm
15 cm 15 cm

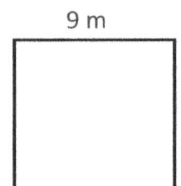
9 m

5) (regular hexagon) *96 m*   6) *56 m*   7) (parallelogram) *28 cm*   8) (regular hexagon) *120 ft*

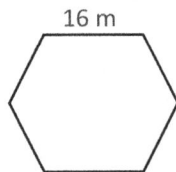
16 m

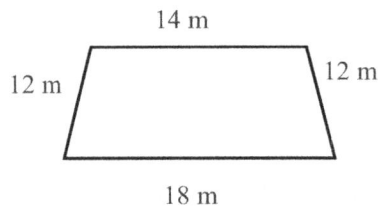
14 m
12 m   12 m
18 m

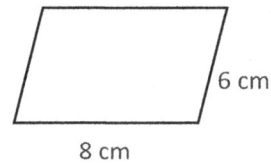
6 cm
8 cm

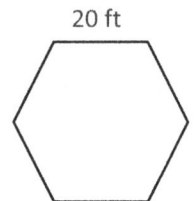
20 ft

9) *72 ft*   10) *72 in*   11) *88 ft*   12) (regular hexagon) *192 in*

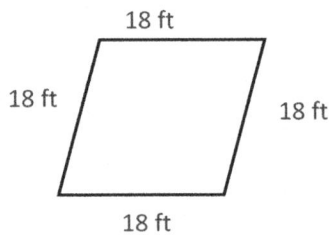
18 ft
18 ft   18 ft
18 ft

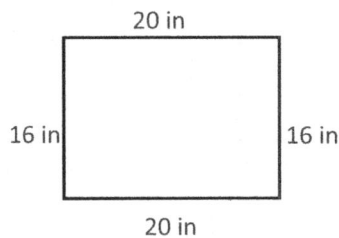
20 in
16 in   16 in
20 in

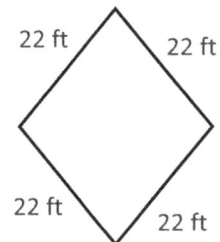
22 ft   22 ft
22 ft   22 ft

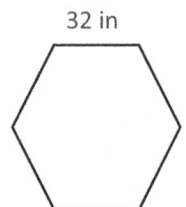
32 in

EffortlessMath.com

Circles

✍ *Find the Circumference of each circle.* (π = 3.14)

1) ____ 2) ____ 3) ____ 4) ____ 5) ____ 6) ____

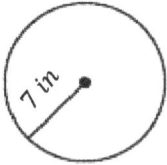 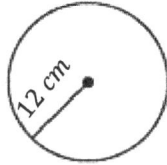 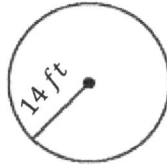 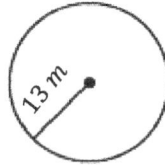 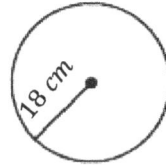 

7) ____ 8) ____ 9) ____ 10) ____ 11) ____ 12) ____

 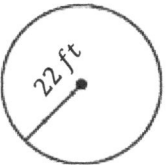 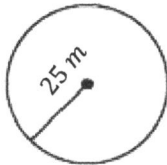 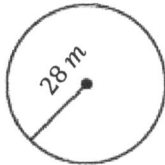 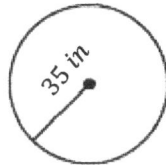 

✍ *Complete the table below.* (π = 3.14)

	Radius	Diameter	Circumference	Area
Circle 1	2 inches	4 inches	12.56 inches	12.56 square inches
Circle 2		8 meters		
Circle 3				113.04 square ft
Circle 4			50.24 miles	
Circle 5		9 km		
Circle 6	7 cm			
Circle 7		10 feet		
Circle 8				615.44 square meters
Circle 9			81.64 inches	
Circle 10	12 feet			

Circles - Answers

✎ **Find the Circumference of each circle.** (π = 3.14)

1) 43.96 *in*	2) 75.36 *cm*	3) 87.92 *ft*	4) 81.64 *m*	5) 113.04 *cm*	6) 94.2 *miles*
7 in	12 cm	14 ft	13 m	18 cm	15 miles
7) 119.32 *in*	8) 138.16 *ft*	9) 157 *m*	10) 175.84 *m*	11) 219.8 *in*	12) 314 *ft*
19 in	22 ft	25 m	28 m	35 in	50 ft

✎ **Complete the table below.** (π = 3.14)

	Radius	Diameter	Circumference	Area
Circle 1	2 inches	4 inches	12.56 inches	12.56 *square inches*
Circle 2	4 *meters*	8 *meters*	25.12 *meters*	50.24 *square meters*
Circle 3	6 *ft*	12 *ft*	37.68	113.04 *square ft*
Circle 4	8 *miles*	16 *miles*	50.24 *miles*	200.96 *square miles*
Circle 5	4.5 *km*	9 *km*	28.26 *km*	63.585 *square km*
Circle 6	7 *cm*	14 *cm*	43.96 *cm*	153.86 *square cm*
Circle 7	5 *feet*	10 *feet*	31.4 *feet*	78.5 *square feet*
Circle 8	14 *m*	28 *m*	87.92 *m*	615.44 *square meters*
Circle 9	13 *in*	26 *in*	81.64 *inches*	530.66 *square inches*
Circle 10	12 *feet*	24 *feet*	75.36 *feet*	452.16 *square feet*

Cubes

✎ **Find the volume of each cube.**

1)
_____
4.5 cm

2)
_____
6.2 cm

3)
_____
9.5 ft

4)
_____
11 m

5)
_____
13 in

6)
_____
8.5 m

7)
_____
7.5 km

8)
_____
6.5 cm

9)
_____
16 ft

10)
_____
17 cm

11)
_____
30 in

12)
_____
40 km

✎ **Find the surface area of each cube.**

13)
_____
25 m

14)
_____
40 m

15)
_____
50 ft

16)
_____
70 mm

17)
_____
60 km

18)
_____
80 cm

Cubes - Answers

✎ **Find the volume of each cube.**

1) $91.125\ cm^3$	2) $238.328\ cm^3$	3) $857.375\ ft^3$	4) $1{,}331\ m^3$	5) $2{,}197\ in^3$	6) $614.125\ m^3$
4.5 cm	6.2 cm	9.5 ft	11 m	13 in	8.5 m

7) $421.875\ km^3$	8) $274.625\ cm^3$	9) $4{,}096\ ft^3$	10) $4{,}913\ cm^3$	11) $27{,}000\ in^3$	12) $64{,}000\ km^3$
7.5 km	6.5 cm	16 ft	17 cm	30 in	40 km

✎ **Find the surface area of each cube.**

13) $3{,}750\ m^2$	14) $9{,}600\ m^2$	15) $15{,}000\ ft^2$	16) $29{,}400\ mm^2$	17) $21{,}600\ km^2$	18) $38{,}400\ cm^2$
25 m	40 m	50 ft	70 mm	60 km	80 cm

EffortlessMath.com

Trapezoids

✎ *Find the area of each trapezoid.*

1) _____

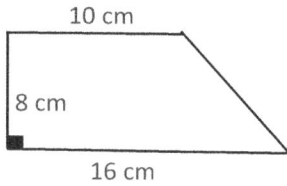

10 cm

8 cm

16 cm

2) _____

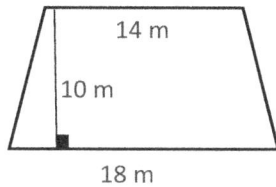

14 m

10 m

18 m

3) _____

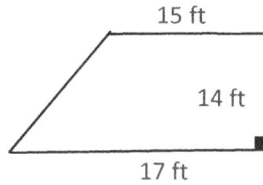

15 ft

14 ft

17 ft

4) _____

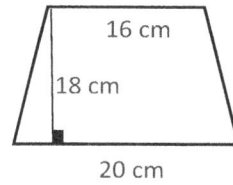

16 cm

18 cm

20 cm

5) _____

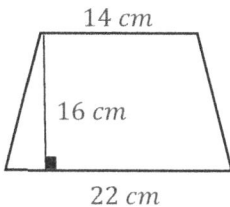

14 cm

16 cm

22 cm

6) _____

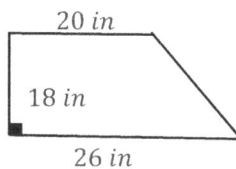

20 in

18 in

26 in

7) _____

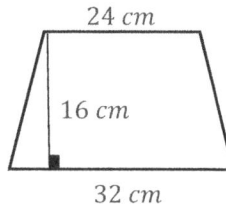

24 cm

16 cm

32 cm

8) _____

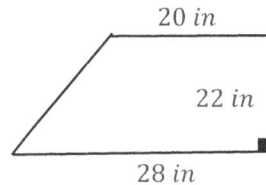

20 in

22 in

28 in

✎ **Solve.**

9) A trapezoid has an area of 80 $cm2$ and its height is 8 cm and one base is 12 cm. What is the other base length? _____

10) If a trapezoid has an area of 120 $ft2$ and the lengths of the bases are 14 ft and 16 ft, find the height. _____

11) If a trapezoid has an area of 160 $m2$ and its height is 10 m and one base is 14 m, find the other base length. _____

12) The area of a trapezoid is 504 $ft2$ and its height is 24 ft. If one base of the trapezoid is 14 ft, what is the other base length?

_____

bit.ly/3hpKACJ

Find more at

Trapezoids - Answers

✎ *Find the area of each trapezoid.*

1) **104 cm^2**

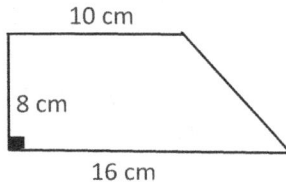

10 cm

8 cm

16 cm

2) **160 m^2**

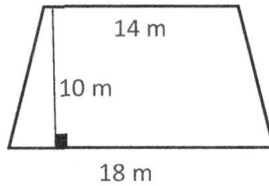

14 m

10 m

18 m

3) **224 ft^2**

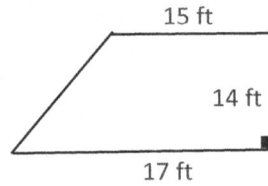

15 ft

14 ft

17 ft

4) **324 cm^2**

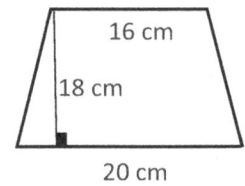

16 cm

18 cm

20 cm

5) **288 cm^2**

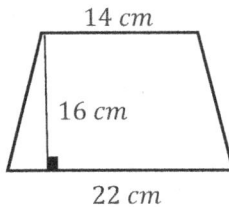

14 cm

16 cm

22 cm

6) **414 in^2**

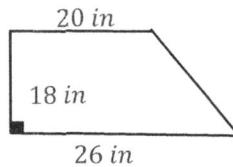

20 in

18 in

26 in

7) **448 cm^2**

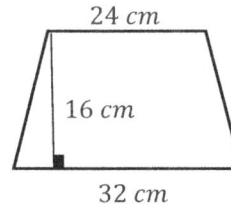

24 cm

16 cm

32 cm

8) **528 in^2**

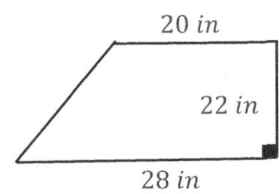

20 in

22 in

28 in

✎ **Solve.**

9) A trapezoid has an area of 80 $cm2$ and its height is 8 cm and one base is 12 cm. What is the other base length? 8 cm

10) If a trapezoid has an area of 120 $ft2$ and the lengths of the bases are14 ft and 16ft, find the height. 8 ft

11) If a trapezoid has an area of 160 $m2$ and its height is 10 m and one base is 14 m, find the other base length. 18 m

12) The area of a trapezoid is 504 $ft2$ and its height is 24 ft. If one base of the trapezoid is 14 ft, what is the other base length? 28 ft

Rectangular Prisms

✍ *Find the volume of each Rectangular Prism.*

1) ____

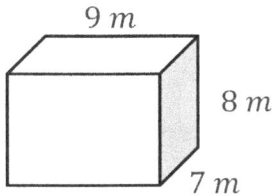

9 m
8 m
7 m

2) ____

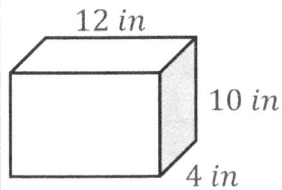

12 in
10 in
4 in

3) ____

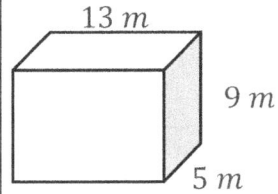

13 m
9 m
5 m

4) ____

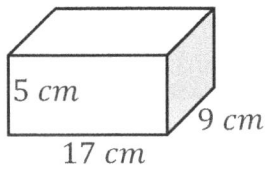

5 cm
17 cm
9 cm

5) ____

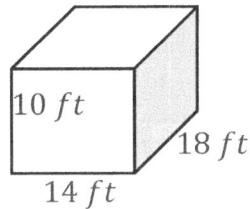

10 ft
18 ft
14 ft

6) ____

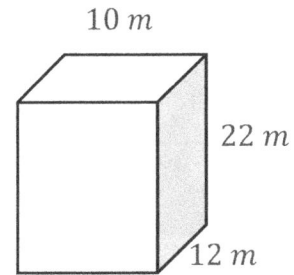

10 m
22 m
12 m

✍ *Find the surface area of each Rectangular Prism.*

7) ____

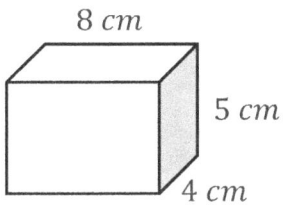

8 cm
5 cm
4 cm

8) ____

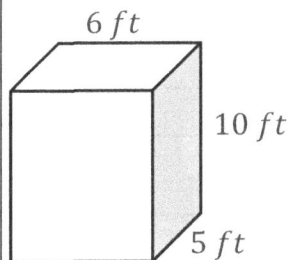

6 ft
10 ft
5 ft

9) ____

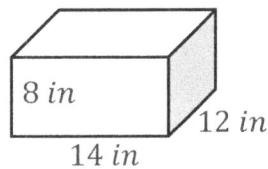

8 in
14 in
12 in

10) ____

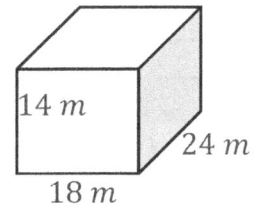

14 m
18 m
24 m

EffortlessMath.com

Rectangular Prisms - Answers

✎ **Find the volume of each Rectangular Prism.**

1) $504\ m^3$

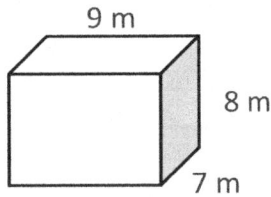

9 m
8 m
7 m

2) $480\ in^3$

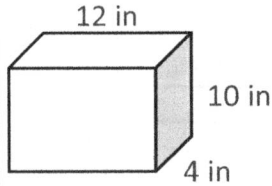

12 in
10 in
4 in

3) $585\ m^3$

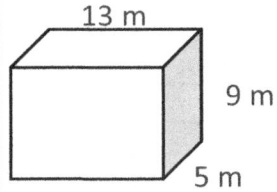

13 m
9 m
5 m

4) $765\ cm^3$

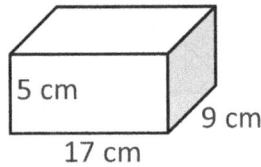

5 cm
9 cm
17 cm

5) $2,520\ ft^3$

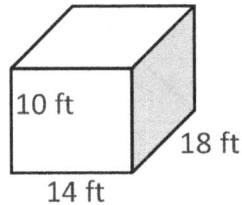

10 ft
18 ft
14 ft

6) $2,640\ m^3$

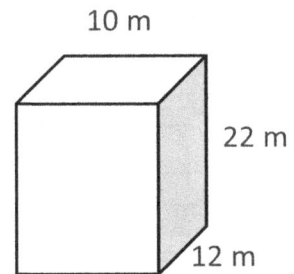

10 m
22 m
12 m

✎ **Find the surface area of each Rectangular Prism.**

7) $184\ cm^2$

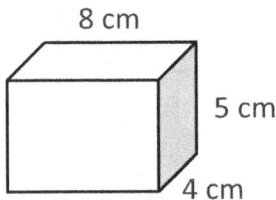

8 cm
5 cm
4 cm

8) $280\ ft^2$

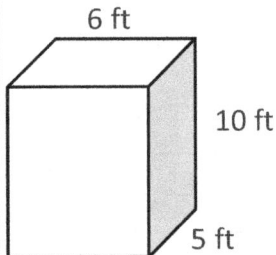

6 ft
10 ft
5 ft

9) $752\ in^2$

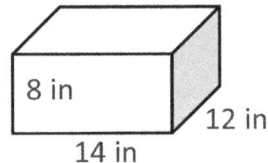

8 in
12 in
14 in

10) $2,040\ m^2$

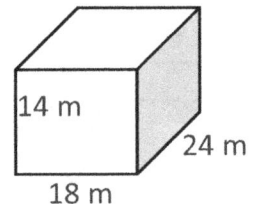

14 m
24 m
18 m

Find more at

bit.ly/3nKm2GT

Cylinder

✎ **Find the volume of each Cylinder.** (π = 3.14)

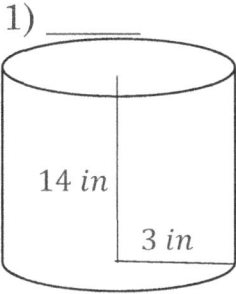

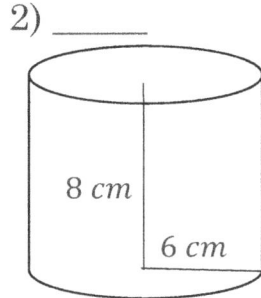

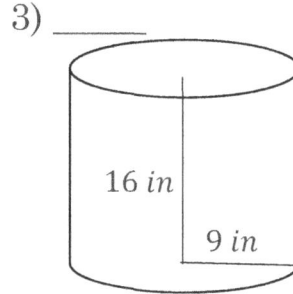

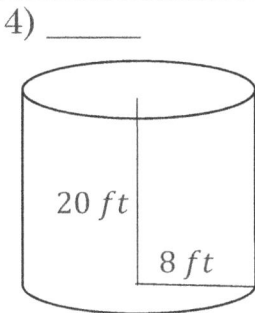

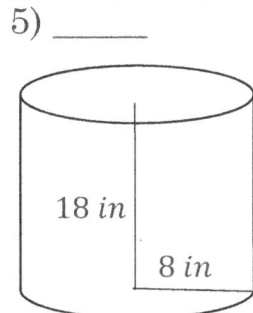

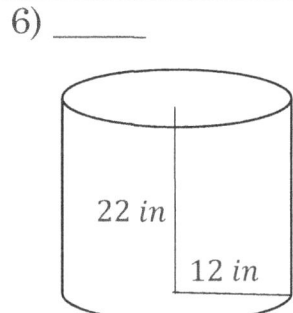

1) _____

14 in
3 in

2) _____

8 cm
6 cm

3) _____

16 in
9 in

4) _____

20 ft
8 ft

5) _____

18 in
8 in

6) _____

22 in
12 in

✎ **Find the surface area of each Cylinder.** (π = 3.14)

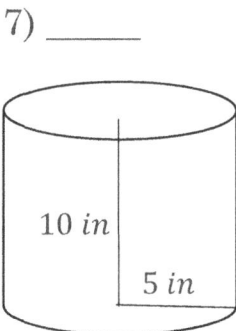

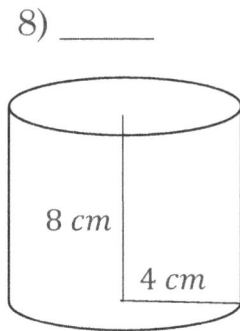

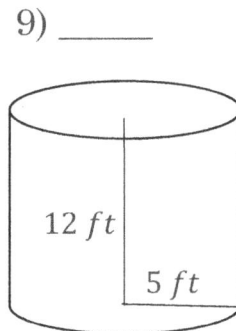

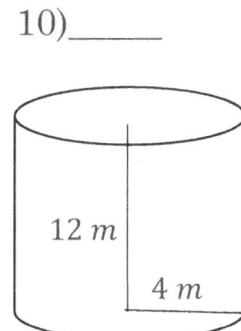

7) _____

10 in
5 in

8) _____

8 cm
4 cm

9) _____

12 ft
5 ft

10) _____

12 m
4 m

Cylinder - Answers

✎ *Find the volume of each Cylinder.* (π = 3.14)

1) **395.64 in³**

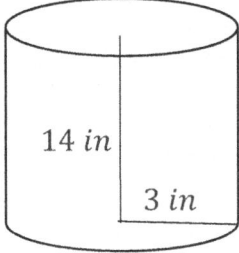

14 in

3 in

2) **904.32 cm³**

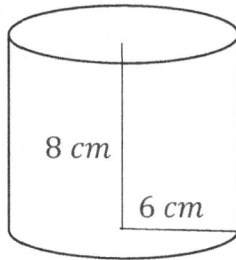

8 cm

6 cm

3) **4,069.44 in³**

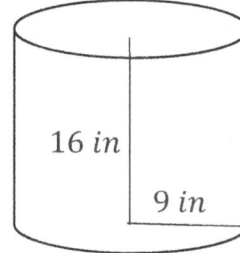

16 in

9 in

4) **4,019.2 ft³**

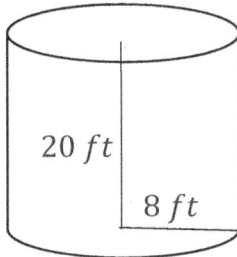

20 ft

8 ft

5) 3,617.28 in³

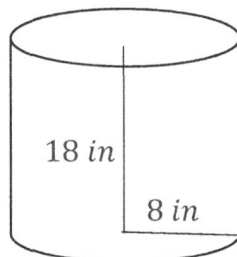

18 in

8 in

6) 9,947.52 in³

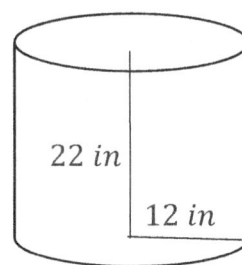

22 in

12 in

✎ *Find the surface area of each Cylinder.* (π = 3.14)

7) 471 in²

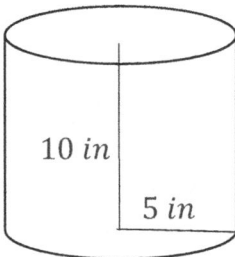

10 in

5 in

8) 301.44 cm²

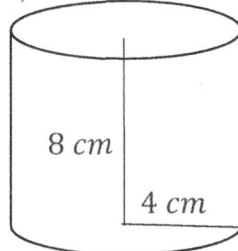

8 cm

4 cm

9) 533.8 ft²

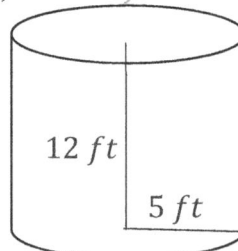

12 ft

5 ft

10) 502.4 m²

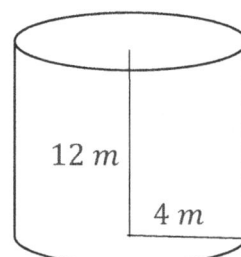

12 m

4 m

Mean, Median, Mode, and Range of the Given Data

✎ *Find the values of the Given Data.*

1) 6, 12, 1, 1, 5

   Mode: _____ Range: _____

   Mean: _____ Median: _____

2) 5, 8, 3, 7, 4, 3

   Mode: _____ Range: _____

   Mean: _____ Median: _____

3) 12, 5, 8, 7, 8

   Mode: _____ Range: _____

   Mean: _____ Median: _____

4) 8, 4, 10, 7, 3, 4

   Mode: _____ Range: _____

   Mean: _____ Median: _____

5) 9, 7, 10, 5, 7, 4, 14

   Mode: _____ Range: _____

   Mean: _____ Median: _____

6) 8, 1, 6, 6, 9, 2, 17

   Mode: _____ Range: _____

   Mean: _____ Median: _____

7) 12, 6, 1, 7, 9, 7, 8, 14

   Mode: _____ Range: _____

   Mean: _____ Median: _____

8) 10, 14, 5, 4, 11, 6, 13

   Mode: _____ Range: _____

   Mean: _____ Median: _____

9) 16, 15, 15, 16, 13, 14, 23

   Mode: _____ Range: _____

   Mean: _____ Median: _____

10) 16, 15, 12, 8, 4, 9, 8, 16

   Mode: _____ Range: _____

   Mean: _____ Median: _____

Mean, Median, Mode, and Range of the Given Data – Answers

✏️ *Find the values of the Given Data.*

1) 6, 12, 1, 1, 5

    Mode: 1   Range: 11

    Mean: 5   Median: 5

2) 5, 8, 3, 7, 4, 3

    Mode: 3   Range: 5

    Mean: 5   Median: 4.5

3) 12, 5, 8, 7, 8

    Mode: 8   Range: 7

    Mean: 8   Median: 8

4) 8, 4, 10, 7, 3, 4

    Mode: 4   Range: 7

    Mean: 6   Median: 5.5

5) 9, 7, 10, 5, 7, 4, 14

    Mode: 7   Range: 10

    Mean: 8   Median: 7

6) 8, 1, 6, 6, 9, 2, 17

    Mode: 6   Range: 16

    Mean: 7   Median: 6

7) 12, 6, 1, 7, 9, 7, 8, 14

    Mode: 7   Range: 13

    Mean: 8   Median: 7.5

8) 10, 14, 5, 4, 11, 6, 13

    Mode: *no mode*   Range: 10

    Mean: 9   Median: 10

9) 16, 15, 15, 16, 13, 14, 23

    Mode: 15 *and* 16 Range: 10

    Mean: 16   Median: 15

10) 16, 15, 12, 8, 4, 9, 8, 16

    Mode: 8 *and* 16   Range: 12

    Mean: 11   Median: 10.5

Pie Graph

✍ *The circle graph below shows all Wilson's expenses for last month.*
Wilson spent $200 on his bills last month.

Wilson's last month expenses

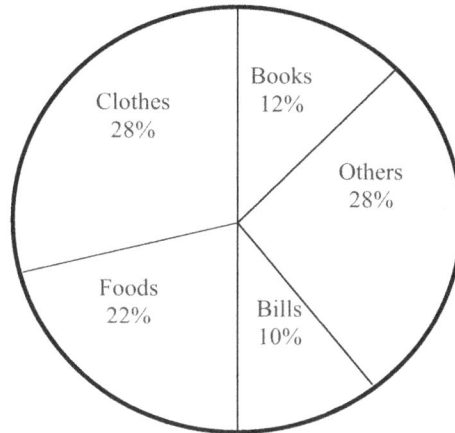

Clothes
28%

Books
12%

Others
28%

Foods
22%

Bills
10%

Answer following questions based on the Pie graph.

1) How much was Wilson's total expenses last month? _____

2) How much did Wilson spend on his clothes last month? _____

3) How much did Wilson spend on foods last month? _____

4) How much did Wilson spend on his books last month? _____

5) What fraction is Wilson's expenses for his bills and clothes out of his total expenses last month? _____

Pie Graph - Answers

✍ *The circle graph below shows all Wilson's expenses for last month.*
Wilson spent $200 on his bills last month.

Wilson's last month expenses

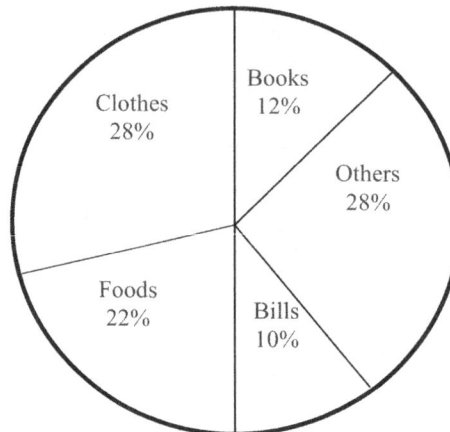

Clothes
28%

Books
12%

Others
28%

Foods
22%

Bills
10%

Answer following questions based on the Pie graph.

1) How much was Wilson's total expenses last month? $2,000

2) How much did Wilson spend on his clothes last month? $560

3) How much did Wilson spend on foods last month? $440

4) How much did Wilson spend on his books last month? $240

5) What fraction is Wilson's expenses for his bills and clothes out of his

total expenses last month? $\dfrac{19}{50}$

Probability Problems

1) If there are 10 red balls and 20 blue balls in a basket, what is the probability that Oliver will pick out a red ball from the basket?

   _____

Gender	Under 45	45 or older	total
Male	12	6	18
Female	5	7	12
Total	17	13	30

2) The table above shows the distribution of age and gender for 30 employees in a company. If one employee is selected at random, what is the probability that the employee selected be either a female under age 45 or a male age 45 or older? _____

3) A number is chosen at random from 1 to 18. Find the probability of not selecting a composite number. (A composite number is a number that is divisible by itself, 1 and at least one other whole number)

   _____

4) There are 6 blue marbles, 8 red marbles, and 5 yellow marbles in a box. If Ava randomly selects a marble from the box, what is the probability of selecting a red or yellow marble? _____

5) A bag contains 19 balls: three green, five black, eight blue, a brown, a red and one white. If 18 balls are removed from the bag at random, what is the probability that a brown ball has been removed?

   _____

6) There are only red and blue marbles in the box. The probability of randomly choosing a red marble in the box is one-fourth. If there are 132 blue marbles, how many are in the box?

   _____

Find more at
bit.ly/3phwk1p

Probability Problems - Answers

1) If there are 10 red balls and 20 blue balls in a basket, what is the probability that Oliver will pick out a red ball from the basket? $\frac{1}{3}$

Gender	Under 45	45 or older	total
Male	12	6	18
Female	5	7	12
Total	17	13	30

2) The table above shows the distribution of age and gender for 30 employees in a company. If one employee is selected at random, what is the probability that the employee selected be either a female under age 45 or a male age 45 or older? $\frac{11}{30}$

3) A number is chosen at random from 1 to 18. Find the probability of not selecting a composite number. (A composite number is a number that is divisible by itself, 1 and at least one other whole number) $\frac{7}{18}$

4) There are 6 blue marbles, 8 red marbles, and 5 yellow marbles in a box. If Ava randomly selects a marble from the box, what is the probability of selecting a red or yellow marble? $\frac{13}{19}$

5) A bag contains 19 balls: three green, five black, eight blue, a brown, a red and one white. If 18 balls are removed from the bag at random, what is the probability that a brown ball has been removed? $\frac{18}{19}$

6) There are only red and blue marbles in the box. The probability of randomly choosing a red marble in the box is one-fourth. If there are 132 blue marbles, how many are in the box? 176

Find more at

bit.ly/3phwk1p

Permutations and Combinations

✎ Calculate the value of each.

1) $5! =$ ____

2) $6! =$ ____

3) $8! =$ ____

4) $5! + 6! =$ ____

5) $8! + 3! =$ ____

6) $6! + 7! =$ ____

7) $8! + 4! =$ ____

8) $9! - 3! =$ ____

✎ Solve each word problems.

9) Sophia is baking cookies. She uses milk, flour and eggs. How many different orders of ingredients can she try? _____

10) William is planning for his vacation. He wants to go to the restaurant, watch a movie, go to the beach, and play basketball. How many different ways of ordering are there for him? _____

11) How many 7-digit numbers can be named using the digits $1, 2, 3, 4, 5, 6$ and 7 without repetition? _____

12) In how many ways can 9 boys be arranged in a straight line? _____

13) In how many ways can 10 athletes be arranged in a straight line? _____

14) A professor is going to arrange her 7 students in a straight line. In how many ways can she do this? _____

15) How many code symbols can be formed with the letters for the word BLACK? _____

16) In how many ways a team of 7 basketball players can choose a captain and co-captain? _____

Permutations and Combinations - Answers

✍ *Calculate the value of each.*

1) $5! = 120$

2) $6! = 720$

3) $8! = 40,320$

4) $5! + 6! = 840$

5) $8! + 3! = 40,326$

6) $6! + 7! = 5,760$

7) $8! + 4! = 40,344$

8) $9! - 3! = 362,874$

✍ *Solve each word problems.*

9) Sophia is baking cookies. She uses milk, flour and eggs. How many different orders of ingredients can she try? 6

10) William is planning for his vacation. He wants to go to the restaurant, watch a movie, go to the beach, and play basketball. How many different ways of ordering are there for him? 24

11) How many 7-digit numbers can be named using the digits $1, 2, 3, 4, 5, 6$ and 7 without repetition? 5,040

12) In how many ways can 9 boys be arranged in a straight line? 362,880

13) In how many ways can 10 athletes be arranged in a straight line? 3,628,800

14) A professor is going to arrange her 7 students in a straight line. In how many ways can she do this? 5,040

15) How many code symbols can be formed with the letters for the word BLACK? 120

16) In how many ways a team of 7 basketball players can choose a captain and co-captain? 42

Find more at

bit.ly/34BQgUY

Function Notation and Evaluation

✎ *Evaluate each function.*

1) $f(x) = x - 1$, find $f(-1)$

_____

2) $g(x) = x + 3$, find $g(4)$

_____

3) $h(x) = x + 9$, find $h(3)$

_____

4) $f(x) = -x - 6$, find $f(5)$

_____

5) $f(x) = 2x - 7$, find $f(-1)$

_____

6) $w(x) = -2 - 4x$, find $w(5)$

_____

7) $g(n) = 6n - 3$, find $g(-2)$

_____

8) $h(x) = -8x + 12$, find $h(3)$

_____

9) $k(n) = 14 - 3n$, find $k(3)$

_____

10) $g(x) = 4x - 4$, find $g(-2)$

_____

11) $k(n) = 8n - 7$, find $k(4)$

_____

12) $w(n) = -2n + 14$, find $w(5)$

_____

13) $h(x) = 5x - 18$, find $h(8)$

_____

14) $g(n) = 2n^2 + 2$, find $g(5)$

_____

15) $f(x) = 3x^2 - 13$, find $f(2)$

_____

16) $g(n) = 5n^2 + 7$, find $g(-3)$

_____

17) $h(n) = 5n^2 - 10$, find $h(4)$

_____

18) $g(x) = -3x^2 - 6x$, find $g(2)$

_____

19) $k(n) = 3n^3 + 2n$, find $k(-5)$

_____

20) $f(x) = -4x + 12$, find $f(2x)$

_____

21) $k(a) = 6a + 5$, find $k(a - 1)$

_____

22) $h(x) = 9x + 3$, find $h(5x)$

_____

Function Notation and Evaluation – Answer

✎ *Evaluate each function.*

1) $f(x) = x - 1$, find $f(-1)$

-2

2) $g(x) = x + 3$, find $g(4)$

7

3) $h(x) = x + 9$, find $h(3)$

12

4) $f(x) = -x - 6$, find $f(5)$

-11

5) $f(x) = 2x - 7$, find $f(-1)$

-9

6) $w(x) = -2 - 4x$, find $w(5)$

-22

7) $g(n) = 6n - 3$, find $g(-2)$

-15

8) $h(x) = -8x + 12$, find $h(3)$

-12

9) $k(n) = 14 - 3n$, find $k(3)$

5

10) $g(x) = 4x - 4$, find $g(-2)$

-12

11) $k(n) = 8n - 7$, find $k(4)$

25

12) $w(n) = -2n + 14$, find $w(5)$

4

13) $h(x) = 5x - 18$, find $h(8)$

22

14) $g(n) = 2n^2 + 2$, find $g(5)$

52

15) $f(x) = 3x^2 - 13$, find $f(2)$

-1

16) $g(n) = 5n^2 + 7$, find $g(-3)$

52

17) $h(n) = 5n^2 - 10$, find $h(4)$

70

18) $g(x) = -3x^2 - 6x$, find $g(2)$

-24

19) $k(n) = 3n^3 + 2n$, find $k(-5)$

-385

20) $f(x) = -4x + 12$, find $f(2x)$

$-8x + 12$

21) $k(a) = 6a + 5$, find $k(a - 1)$

$6a - 1$

22) $h(x) = 9x + 3$, find $h(5x)$

$45x + 3$

Adding and Subtracting Functions

✎ *Perform the indicated operation.*

1) $f(x) = x + 6$

   $g(x) = 3x + 3$

   Find $(f - g)(2)$

   _____

2) $g(x) = x - 3$

   $f(x) = -x - 4$

   Find $(g - f)(-2)$

   _____

3) $h(t) = 5t + 4$

   $g(t) = 2t + 2$

   Find $(h + g)(-1)$

   _____

4) $g(a) = 3a - 5$

   $f(a) = a^2 + 6$

   Find $(g + f)(3)$

   _____

5) $g(x) = 4x - 5$

   $h(x) = 6x^2 + 5$

   Find $(g - f)(-2)$

   _____

6) $h(x) = x^2 + 3$

   $g(x) = -4x + 1$

   Find $(h + g)(4)$

   _____

7) $f(x) = -2x - 8$

   $g(x) = x^2 + 2$

   Find $(f - g)(6)$

   _____

8) $h(n) = -4n^2 + 9$

   $g(n) = 5n + 6$

   Find $(h - g)(5)$

   _____

9) $g(x) = 3x^2 - 2x - 1$

   $f(x) = 5x + 12$

   Find $(g - f)(a)$

   _____

10) $g(t) = -5t - 8$

    $f(t) = -t^2 + 2t + 12$

    Find $(g + f)(x)$

    _____

Adding and Subtracting Functions - answer

✎ *Perform the indicated operation.*

1) $f(x) = x + 6$

   $g(x) = 3x + 3$

   Find $(f - g)(2)$

   -1

2) $g(x) = x - 3$

   $f(x) = -x - 4$

   Find $(g - f)(-2)$

   -3

3) $h(t) = 5t + 4$

   $g(t) = 2t + 2$

   Find $(h + g)(-1)$

   -1

4) $g(a) = 3a - 5$

   $f(a) = a^2 + 6$

   Find $(g + f)(3)$

   19

5) $g(x) = 4x - 5$

   $h(x) = 6x^2 + 5$

   Find $(g - f)(-2)$

   -42

6) $h(x) = x^2 + 3$

   $g(x) = -4x + 1$

   Find $(h + g)(4)$

   4

7) $f(x) = -2x - 8$

   $g(x) = x^2 + 2$

   Find $(f - g)(6)$

   -58

8) $h(n) = -4n^2 + 9$

   $g(n) = 5n + 6$

   Find $(h - g)(5)$

   -122

9) $g(x) = 3x^2 - 2x - 1$

   $f(x) = 5x + 12$

   Find $(g - f)(a)$

   $3a^2 - 7a - 13$

10) $g(t) = -5t - 8$

    $f(t) = -t^2 + 2t + 12$

    Find $(g + f)(x)$

    $-x^2 - 3x + 4$

bit.ly/3hdeFVO

Multiplying and Dividing Functions

✎ *Perform the indicated operation.*

1) $g(x) = x + 2$

   $f(x) = x + 3$

   Find $(g.f)(4)$

   _____

2) $g(a) = a + 2$

   $h(a) = 2a - 3$

   Find $(g.h)(5)$

   _____

3) $f(x) = a^2 - 2$

   $g(x) = -4 + 3a$

   Find $(fg)(2)$

   _____

4) $g(t) = t^2 + 4$

   $h(t) = 2t - 4$

   Find $(g.h)(-3)$

   _____

5) $g(a) = 2a^2 - 4a + 2$

   $f(a) = 2a^3 - 2$

   Find $(\frac{g}{f})(4)$

   _____

6) $f(x) = 2x$

   $h(x) = -x + 6$

   Find $(f.h)(-2)$

   _____

7) $f(x) = 2x + 4$

   $h(x) = 4x - 2$

   Find $(\frac{f}{h})(2)$

   _____

8) $g(a) = 4a + 6$

   $f(a) = 2a - 8$

   Find $(\frac{g}{f})(3)$

   _____

9) $g(x) = x^2 + 2x + 5$

   $h(x) = 2x + 3$

   Find $(g.h)(2)$

   _____

10) $g(x) = -4x^2 + 5 - 2x$

    $f(x) = x^2 - 2$

    Find $(g.f)(3)$

    _____

Multiplying and Dividing Functions - answer

✎ *Perform the indicated operation.*

1) $g(x) = x + 2$

   $f(x) = x + 3$

   Find $(g.f)(4)$

   42

2) $g(a) = a + 2$

   $h(a) = 2a - 3$

   Find $(g.h)(5)$

   49

3) $f(x) = a^2 - 2$

   $g(x) = -4 + 3a$

   Find $(fg)(2)$

   1

4) $g(t) = t^2 + 4$

   $h(t) = 2t - 4$

   Find $(g.h)(-3)$

   -130

5) $g(a) = 2a^2 - 4a + 2$

   $f(a) = 2a^3 - 2$

   Find $(\frac{g}{f})(4)$

   $\frac{1}{7}$

6) $f(x) = 2x$

   $h(x) = -x + 6$

   Find $(f.h)(-2)$

   -32

7) $f(x) = 2x + 4$

   $h(x) = 4x - 2$

   Find $(\frac{f}{h})(2)$

   $\frac{4}{3}$

8) $g(a) = 4a + 6$

   $f(a) = 2a - 8$

   Find $(\frac{g}{f})(3)$

   -9

9) $g(x) = x^2 + 2x + 5$

   $h(x) = 2x + 3$

   Find $(g.h)(2)$

   91

10) $g(x) = -4x^2 + 5 - 2x$

   $f(x) = x^2 - 2$

   Find $(g.f)(3)$

   -259

Composition of Functions

Using $f(x) = x + 4$ **and** $g(x) = 2x$, **find:**

1) $f\big(g(1)\big) = $ _____

2) $f\big(g(-1)\big) = $ _____

3) $g\big(f(-2)\big) = $ _____

4) $g\big(f(2)\big) = $ _____

5) $f\big(g(2)\big) = $ _____

6) $g\big(f(3)\big) = $ _____

Using $f(x) = 2x + 5$ **and** $g(x) = x - 2$, **find:**

7) $g\big(f(2)\big) = $ _____

8) $g\big(f(-2)\big) = $ _____

9) $f\big(g(5)\big) = $ _____

10) $f\big(f(4)\big) = $ _____

11) $g\big(f(3)\big) = $ _____

12) $g\big(f(-3)\big) = $ _____

Using $f(x) = 4x - 2$ **and** $g(x) = x - 5$, **find:**

13) $g\big(f(-2)\big) = $ _____

14) $f\big(f(4)\big) = $ _____

15) $f\big(g(5)\big) = $ _____

16) $f\big(f(3)\big) = $ _____

17) $g\big(f(-3)\big) = $ _____

18) $g\big(g(6)\big) = $ _____

Using $f(x) = 5x + 3$ **and** $g(x) = 2x - 5$, **find:**

19) $f\big(g(-4)\big) = $ _____

20) $g\big(f(6)\big) = $ _____

21) $f\big(g(5)\big) = $ _____

22) $f\big(f(3)\big) = $ _____

Composition of Functions - Answer

Using $f(x) = x + 4$ *and* $g(x) = 2x$, *find:*

1) $f(g(1)) = 6$

2) $f(g(-1)) = 2$

3) $g(f(-2)) = 4$

4) $g(f(2)) = 12$

5) $f(g(2)) = 8$

6) $g(f(3)) = 14$

Using $f(x) = 2x + 5$ *and* $g(x) = x - 2$, *find:*

7) $g(f(2)) = 7$

8) $g(f(-2)) = -1$

9) $f(g(5)) = 11$

10) $f(f(4)) = 31$

11) $g(f(3)) = 9$

12) $g(f(-3)) = -3$

Using $f(x) = 4x - 2$ *and* $g(x) = x - 5$, *find:*

13) $g(f(-2)) = -15$

14) $f(f(4)) = 54$

15) $f(g(5)) = -2$

16) $f(f(3)) = 38$

17) $g(f(-3)) = -19$

18) $g(g(6)) = -4$

Using $f(x) = 5x + 3$ *and* $g(x) = 2x - 5$, *find:*

19) $f(g(-4)) = -62$

20) $g(f(6)) = 61$

21) $f(g(5)) = 28$

22) $f(f(3)) = 93$

Function Inverses

✎ *Find the inverse of each function.*

1) $f(x) = \frac{1}{x} - 6 \rightarrow f^{-1}(x) =$

2) $g(x) = \frac{7}{-x-3} \rightarrow g^{-1}(x) =$

3) $h(x) = \frac{x+9}{3} \rightarrow h^{-1}(x) =$

4) $h(x) = \frac{2x-10}{4} \rightarrow h^{-1}(x) =$

5) $f(x) = \frac{-15+x}{3} \rightarrow f^{-1}(x) =$

6) $s(x) = \sqrt{x} - 2 \rightarrow s^{-1}(x) =$

7) $f(x) = -x + 10 \rightarrow i^{-1}(x) =$

8) $g(x) = -7x + 4 \rightarrow g^{-1}(x) =$

9) $f(x) = 5x - 1 \rightarrow f^{-1}(x) =$

10) $h(x) = 3(x-1)^3 \rightarrow h^{-1}(x) =$

11) $f(x) = -(x+2)^3 \rightarrow f^{-1}(x) =$

12) $s(x) = -\sqrt{x} + 7 \rightarrow s^{-1}(x) =$

13) $h(x) = -\sqrt{x} + 8 \rightarrow h^{-1}(x) =$

✎ *Find the inverse. Then graph the function and its inverse.*

14) $f(x) - -2 - \frac{1}{3}x$

15) $g(x) = -2x^3 + 3$

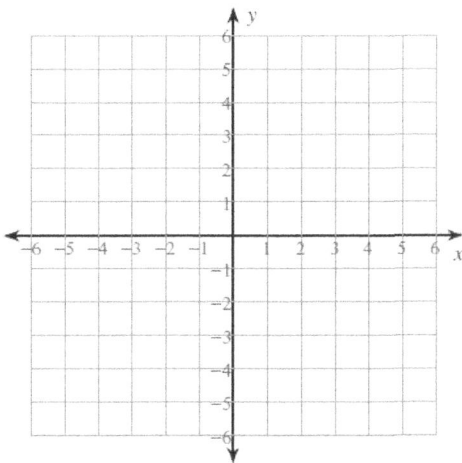

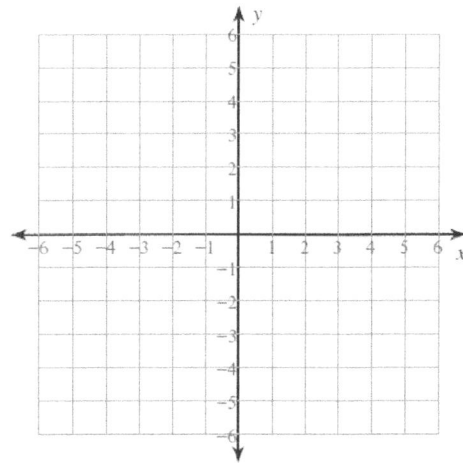

Function Inverses - Answers

✎ *Find the inverse of each function.*

1) $f(x) = \frac{1}{x} - 6 \rightarrow f^{-1}(x) = \frac{1}{x+6}$

2) $g(x) = \frac{7}{-x-3} \rightarrow g^{-1}(x) = -\frac{1+3x}{x}$

3) $h(x) = \frac{x+9}{3} \rightarrow h^{-1}(x) = 3x - 9$

4) $h(x) = \frac{2x-10}{4} \rightarrow h^{-1}(x) = 2x + 5$

5) $f(x) = \frac{-15+x}{3} \rightarrow f^{-1}(x) = 3x + 15$

6) $s(x) = \sqrt{x} - 2 \rightarrow s^{-1}(x) = x^2 + 4x + 4$

7) $f(x) = -x + 10 \rightarrow f^{-1}(x) = 10 - x$

8) $g(x) = -7x + 4 \rightarrow g^{-1}(x) = -\frac{x-4}{7}$

9) $f(x) = 5x - 1 \rightarrow f^{-1}(x) = \frac{x+1}{5}$

10) $h(x) = 3(x-1)^3 \rightarrow h^{-1}(x) = \sqrt[3]{\frac{x}{3}} + 1$

11) $f(x) = -(x+2)^3 \rightarrow f^{-1}(x) = -\sqrt[3]{x} - 2$

12) $s(x) = -\sqrt{x} + 7 \rightarrow s^{-1}(x) = x^2 - 14x + 49$

13) $f(x) = -\sqrt{x} - 8 \rightarrow f^{-1}(x) = x^2 + 16x + 6$

✎ *Find the inverse of each function. Then graph its inverse.*

14) $f(x) = -2 - \frac{1}{3}x$

15) $g(x) = -2x + 3$

$f^{-1}(x) = -3x - 6$

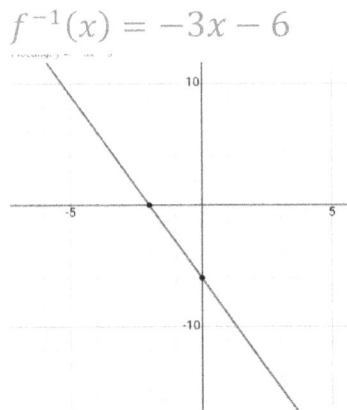

$g^{-1}(x) = -\frac{x-3}{2}$

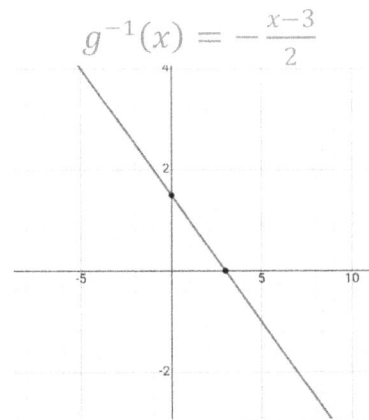

Solving a Quadratic Equation

Solve each equation by factoring or using the quadratic formula.

1) $x^2 - 4x - 32 = 0$

_____

2) $x^2 - 2x - 63 = 0$

_____

3) $x^2 + 17x + 72 = 0$

_____

4) $x^2 + 14x + 48 = 0$

_____

5) $x^2 + 5x - 24 = 0$

_____

6) $x^2 + 15x + 36 = 0$

_____

7) $x^2 + 12x - 28 = 0$

_____

8) $x^2 + 6x - 55 = 0$

_____

9) $x^2 + 16x - 105 = 0$

_____

10) $x^2 - 21x + 54 = 0$

_____

11) $x^2 + 8x - 128 = 0$

_____

12) $x^2 + 19x - 150 = 0$

_____

13) $x^2 + 15x - 154 = 0$

_____

14) $2x^2 - 2x - 60 = 0$

_____

15) $2x^2 - 10x - 72 = 0$

_____

16) $4x^2 + 48x + 128 = 0$

_____

17) $4x^2 + 40x + 96 = 0$

_____

18) $2x^2 + 28x + 90 = 0$

_____

19) $9x^2 + 63x + 108 = 0$

_____

20) $4x^2 + 56x + 160 = 0$

_____

Solving a Quadratic Equation - Answers

✍ *Solve each equation by factoring or using the quadratic formula.*

1) $x^2 - 4x - 32 = 0$

$x = 8, x = -4$

2) $x^2 - 2x - 63 = 0$

$x = 9, x = -7$

3) $x^2 + 17x + 72 = 0$

$x = -9, x = -8$

4) $x^2 + 14x + 48 = 0$

$x = -6, x = -8$

5) $x^2 + 5x - 24 = 0$

$x = 3, x = -8$

6) $x^2 + 15x + 36 = 0$

$x = -12, x = -3$

7) $x^2 + 12x - 28 = 0$

$x = 2, x = -14$

8) $x^2 + 6x - 55 = 0$

$x = 5, x = -11$

9) $x^2 + 16x - 105 = 0$

$x = -21, x = 5$

10) $x^2 - 21x + 54 = 0$

$x = 18, x = 3$

11) $x^2 + 8x - 128 = 0$

$x = -16, x = 8$

12) $x^2 + 19x - 150 = 0$

$x = -25, x = 6$

13) $x^2 + 15x - 154 = 0$

$x = -22, x = 7$

14) $2x^2 - 2x - 60 = 0$

$x = 6, x = -5$

15) $2x^2 - 10x - 72 = 0$

$x = 9, x = -4$

16) $4x^2 + 48x + 128 = 0$

$x = -4, x = -8$

17) $4x^2 + 40x + 96 = 0$

$x = -4, x = -6$

18) $2x^2 + 28x + 90 = 0$

$x = -5, x = -9$

19) $9x^2 + 63x + 108 = 0$

$x = -3, x = -4$

20) $4x^2 + 56x + 160 = 0$

$x = -4, x = -10$

Graphing Quadratic Functions

✎ *Sketch the graph of each function.*

1) $y = (x + 1)^2 - 2$ 2) $y = (x - 1)^2 + 3$ 3) $y = x^2 - 4x + 6$

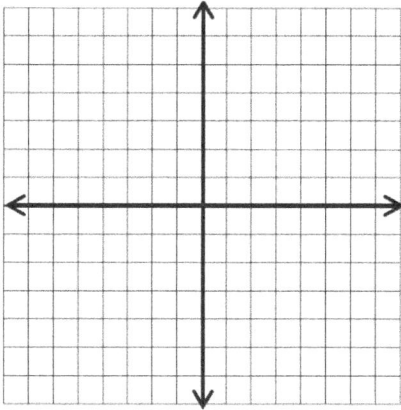   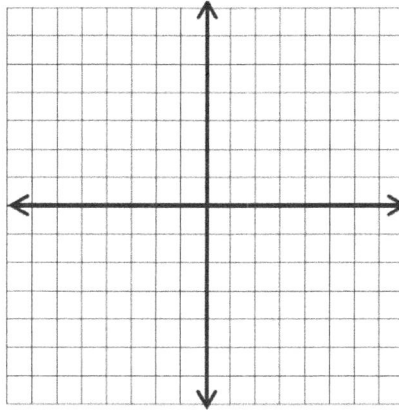   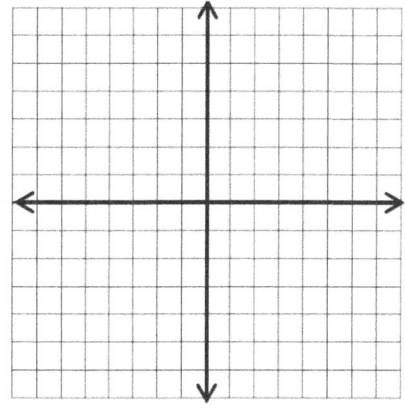

4) $y = x^2 - 6x + 14$ 5) $y = x^2 + 12x + 34$ 6) $y = 2(x + 1)^2 - 4$

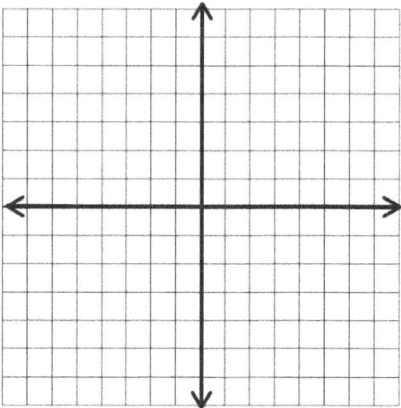   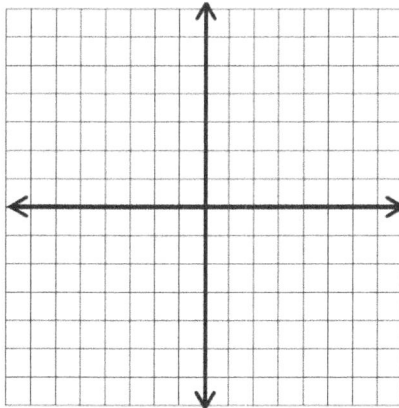   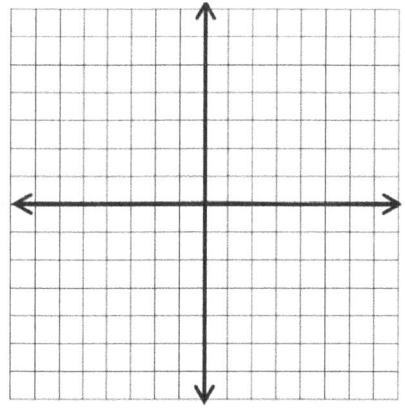

144

Graphing Quadratic Functions - Answers

✎ *Sketch the graph of each function.*

1) $y = (x + 1)^2 - 2$

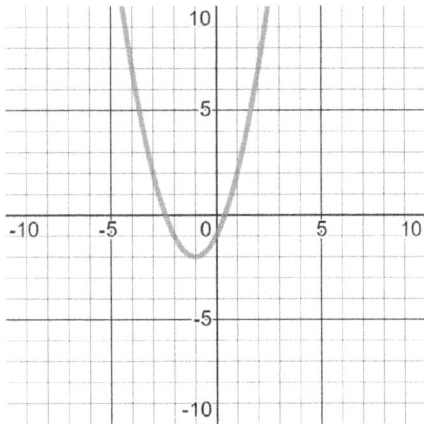

2) $y = (x - 1)^2 + 3$

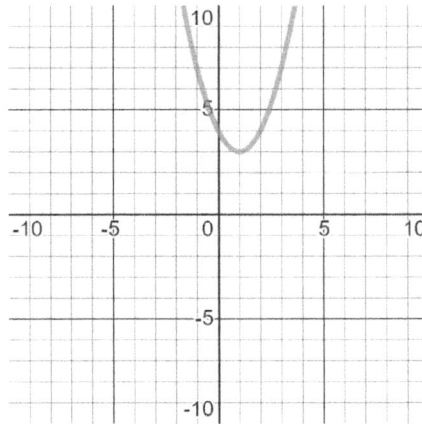

3) $y = x^2 - 4x + 6$

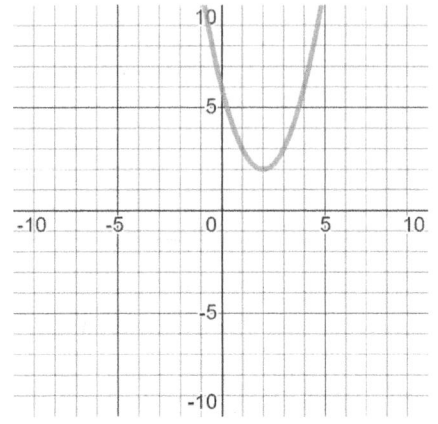

4) $y = x^2 - 6x + 14$

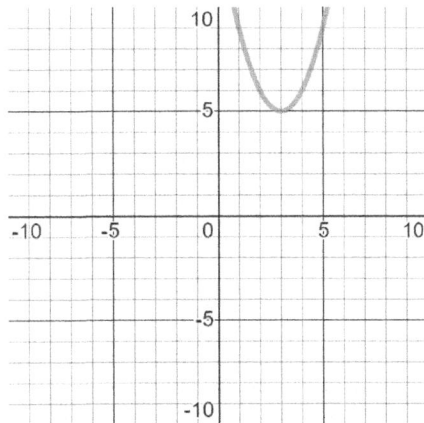

5) $y = x^2 + 12x + 34$

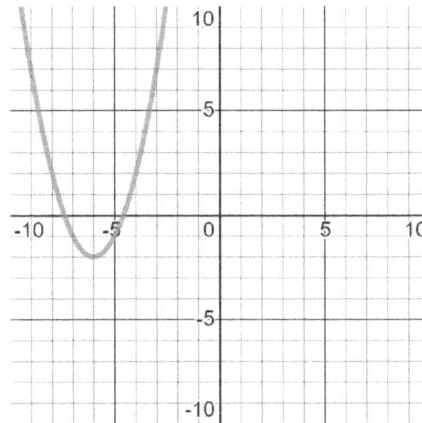

6) $y = 2(x + 1)^2 - 4$

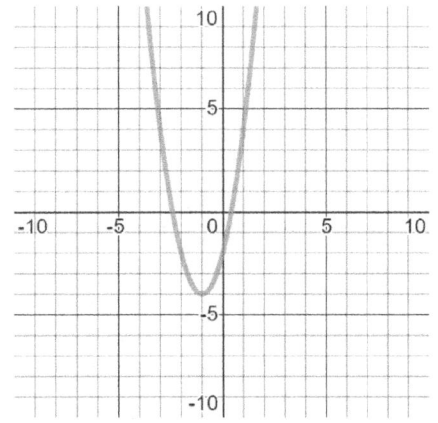

Find more at

bit.ly/3hfFrPB

Solving Quadratic Inequalities

✍ *Solve each quadratic inequality.*

1) $x^2 - 4 < 0$

_____

2) $x^2 - 9 > 0$

_____

3) $x^2 - 5x - 6 < 0$

_____

4) $x^2 + 8x - 20 > 0$

_____

5) $x^2 + 10x - 24 \geq 0$

_____

6) $x^2 - 15x + 54 < 0$

_____

7) $x^2 + 17x + 72 \leq 0$

_____

8) $x^2 + 15x + 44 \geq 0$

_____

9) $x^2 + 5x - 50 \geq 0$

_____

10) $x^2 - 18x + 72 < 0$

_____

11) $x^2 - 18x + 45 > 0$

_____

12) $x^2 + 16x - 80 > 0$

_____

13) $x^2 + 9x - 112 \leq 0$

_____

14) $x^2 + 4x - 117 \leq 0$

_____

15) $x^2 + 19x + 88 \geq 0$

_____

16) $x^2 + 26x + 168 \leq 0$

_____

17) $4x^2 + 24x + 32 < 0$

_____

18) $4x^2 - 4x - 48 \geq 0$

_____

19) $4x^2 - 16x - 48 \leq 0$

_____

20) $9x^2 - 63x + 108 > 0$

_____

Solving Quadratic Inequalities- Answers

✎ *Solve each quadratic inequality.*

1) $x^2 - 4 < 0$

$-2 < x < 2$

2) $x^2 - 9 > 0$

$-3 < x < 3$

3) $x^2 - 5x - 6 < 0$

$-1 < x < 6$

4) $x^2 + 8x - 20 > 0$

$x < -10 \, or \, x > 2$

5) $x^2 + 10x - 24 \geq 0$

$x \leq -12 \, or \, x \geq 2$

6) $x^2 - 15x + 54 < 0$

$6 < x < 9$

7) $x^2 + 17x + 72 \leq 0$

$-9 \leq x \leq -8$

8) $x^2 + 15x + 44 \geq 0$

$x \leq -11 \, or \, x \geq -4$

9) $x^2 + 5x - 50 \geq 0$

$x \leq -10 \, or \, x \geq 5$

10) $x^2 - 18x + 72 < 0$

$6 \leq x \leq 12$

11) $x^2 - 18x + 45 > 0$

$x < 3 \, or \, x > 15$

12) $x^2 + 16x - 80 > 0$

$x < -20 \, or \, x > 4$

13) $x^2 + 9x - 112 \leq 0$

$-16 \leq x \leq 7$

14) $x^2 + 4x - 117 \leq 0$

$-13 \leq x \leq 9$

15) $x^2 + 19x + 88 \geq 0$

$x \leq -11 \, or \, x \geq -8$

16) $x^2 + 26x + 168 \leq 0$

$-14 \leq x \leq -12$

17) $4x^2 + 24x + 32 \leq 0$

$-4 \leq x \leq -2$

18) $4x^2 - 4x - 48 \geq 0$

$x \leq -3 \, or \, x \geq 4$

19) $4x^2 - 16x - 48 \leq 0$

$-2 \leq x \leq 6$

20) $9x^2 - 63x + 108 > 0$

$x \leq 3 \, or \, x \geq 4$

Graphing Quadratic Inequalities

✎ *Sketch the graph of each quadratic inequality.*

1) $y < -2x^2$

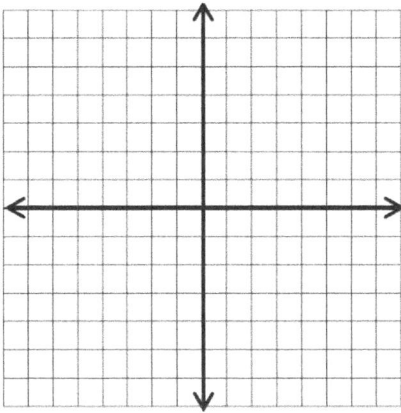

2) $y > 3x^2$

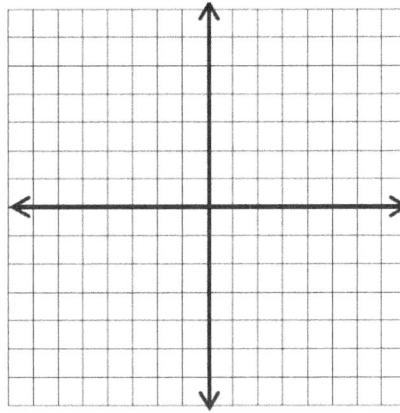

3) $y \geq -3x^2$

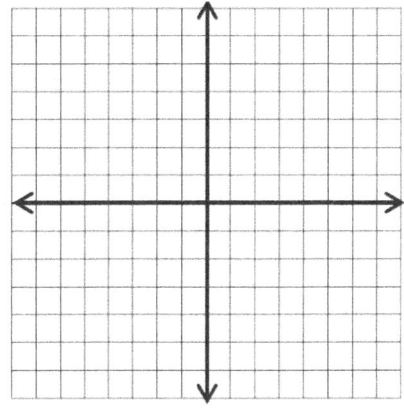

4) $y < x^2 + 1$

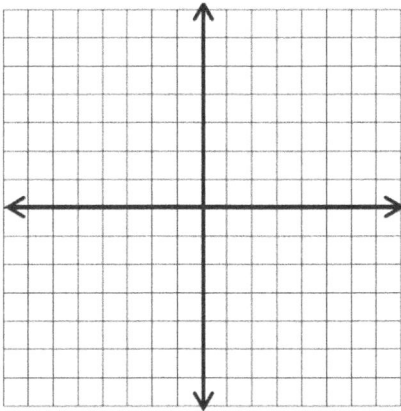

5) $y \geq -x^2 + 2$

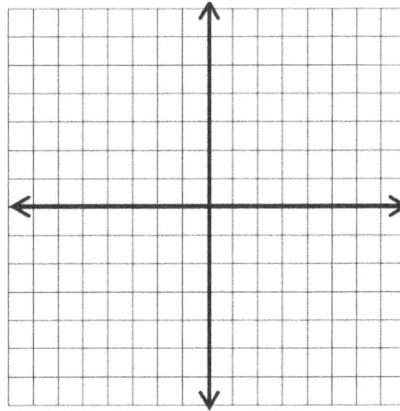

6) $y \leq x^2 - 2x - 3$

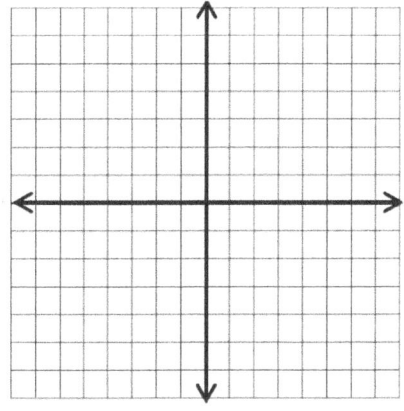

EffortlessMath.com

Find more at

Graphing Quadratic Inequalities- Answers

✎ *Sketch the graph of each quadratic inequality.*

1) $y < -2x^2$

2) $y > 3x^2$

3) $y \geq -3x^2$

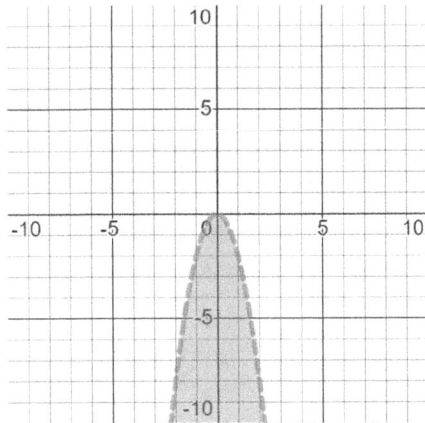

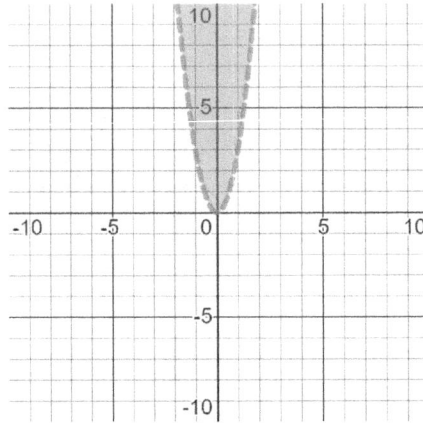

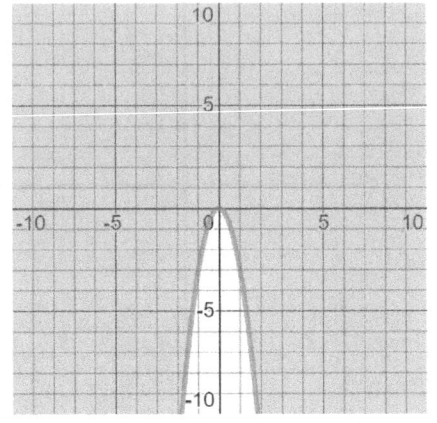

4) $y < x^2 + 1$

5) $y \geq -x^2 + 2$

6) $y \leq x^2 - 2x - 3$

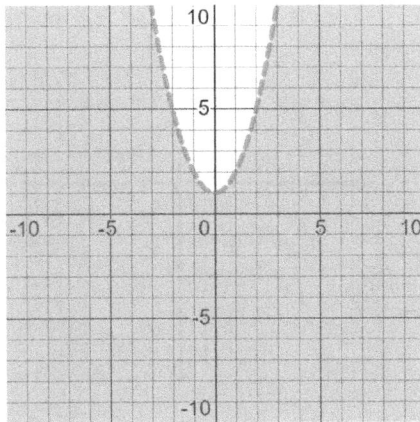

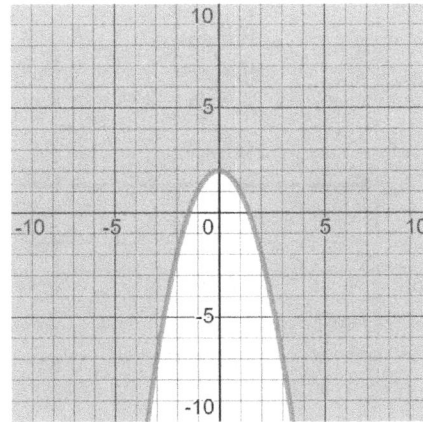

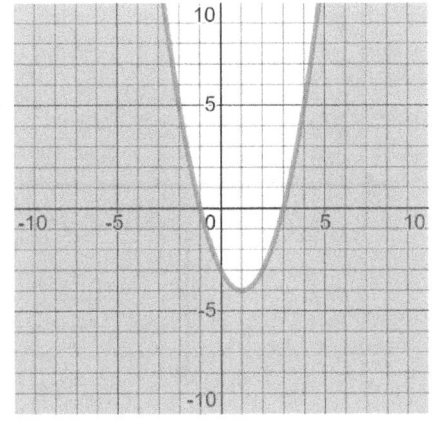

Adding and Subtracting Complex Numbers

✍ *Evaluate.*

1) $(-5i) - (7i) =$

_____

2) $(-2i) + (-8i) =$

_____

3) $(2i) - (6 + 3i) =$

_____

4) $(4 - 6i) + (-2i) =$

_____

5) $(-7i) + (4 + 5i) =$

_____

6) $10 + (-2 - 6i) =$

_____

7) $(-3i) - (9 + 2i) =$

_____

8) $(4 + 6i) - (-3i) =$

_____

9) $(12 + 6i) + (-8i) =$

_____

10) $(14i) - (10 - 5i) =$

_____

11) $(10 + 9i) + (-6i) =$

_____

12) $(11i) - (15 + 3i) =$

_____

13) $(5 + 7i) + (6 + 3i) =$

_____

14) $(9 - 3i) + (5 + i) =$

_____

15) $(12 + 5i) + (6 + 4i) =$

_____

16) $(-2 + 6i) - (-7 - 2i) =$

_____

17) $(-6 + 7i) - (-10 + 3i) =$

_____

18) $(-12 + i) - (-16 - 11i) =$

_____

19) $(-20 - 3i) + (12 + 6i) =$

_____

20) $(-24 - 12i) - (-7 - 9i) =$

_____

Adding and Subtracting Complex Numbers - Answers

✎ *Evaluate.*

1) $(-5i) - (7i) =$

$-12i$

2) $(-2i) + (-8i) =$

$-10i$

3) $(2i) - (6 + 3i) =$

$-6 - i$

4) $(4 - 6i) + (-2i) =$

$4 - 8i$

5) $(-7i) + (4 + 5i) =$

$4 - 2i$

6) $10 + (-2 - 6i) =$

$8 - 6i$

7) $(-3i) - (9 + 2i) =$

$-9 - 5i$

8) $(4 + 6i) - (-3i) =$

$4 + 9i$

9) $(12 + 6i) + (-8i) =$

$12 - 2i$

10) $(14i) - (10 - 5i) =$

$-10 + 19i$

11) $(10 + 9i) + (-6i) =$

$10 + 3i$

12) $(11i) - (15 + 3i) =$

$-15 + 8i$

13) $(5 + 7i) + (6 + 3i) =$

$11 + 10i$

14) $(9 - 3i) + (5 + i) =$

$14 - 2i$

15) $(12 + 5i) + (6 + 4i) =$

$18 + 9i$

16) $(-2 + 6i) - (-7 - 2i) =$

$5 + 8i$

17) $(-6 + 7i) - (-10 + 3i) =$

$4 + 4i$

18) $(-12 + i) - (-16 - 11i) =$

$4 + 12i$

19) $(-20 - 3i) + (12 + 6i) =$

$-8 + 3i$

20) $(-24 - 12i) - (-7 - 9i) =$

$-17 - 3i$

Multiplying and Dividing Complex Numbers

✍ *Calculate.*

1) $(3 - 2i)(4 - 3i) =$

_____

2) $(6 + 2i)(3 + 2i) =$

_____

3) $(8 - i)(4 - 2i) =$

_____

4) $(2 - 4i)(3 - 5i) =$

_____

5) $(5 + 6i)(3 + 2i) =$

_____

6) $(5 + 3i)(9 + 2i) =$

_____

7) $(5 - 8i)(-6 + 3i) =$

_____

8) $\dfrac{4}{-7i} =$

_____

9) $\dfrac{2 - 3i}{3i} =$

_____

10) $\dfrac{6 + 2i}{3i} =$

_____

11) $\dfrac{8i}{-2 + 3i} =$

_____

12) $\dfrac{-4 - 6i}{5i} =$

_____

13) $\dfrac{2 - 9i}{3 - i} =$

_____

14) $\dfrac{8 + 3i}{4 + 4i} =$

_____

15) $\dfrac{8 + 6i}{-2 + 4i} =$

_____

16) $\dfrac{-3 - i}{-6 - 3i} =$

_____

Multiplying and Dividing Complex Numbers - Answers

✎ *Calculate.*

1) $(3 - 2i)(4 - 3i) =$

$6 - 17i$

2) $(6 + 2i)(3 + 2i) =$

$14 + 18i$

3) $(8 - i)(4 - 2i) =$

$30 - 20i$

4) $(2 - 4i)(3 - 5i) =$

$-14 - 22i$

5) $(5 + 6i)(3 + 2i) =$

$3 + 28i$

6) $(5 + 3i)(9 + 2i) =$

$39 + 37i$

7) $(5 - 8i)(-6 + 3i) =$

$-6 + 63i$

8) $\dfrac{4}{-7i} = \dfrac{4}{7}i$

9) $\dfrac{2-3i}{3i} =$

$1 + \dfrac{2}{3}i$

10) $\dfrac{6+2i}{3i} =$

$-\dfrac{2}{3} + 2i$

11) $\dfrac{8i}{-2+3i} =$

$\dfrac{24}{13} - \dfrac{16}{13}i$

12) $\dfrac{-4-6i}{5i} =$

$\dfrac{6}{5} - \dfrac{4}{5}i$

13) $\dfrac{2-9i}{3-i} =$

$\dfrac{3}{2} - \dfrac{5}{2}i$

14) $\dfrac{8+3i}{4+4i} =$

$\dfrac{11}{8} - \dfrac{5}{8}i$

15) $\dfrac{8+6i}{-2+4i} =$

$\dfrac{2}{5} - \dfrac{11}{5}i$

16) $\dfrac{-3-i}{-6-3i} = \dfrac{7}{15} - \dfrac{1}{15}i$

Rationalizing Imaginary Denominators

✍️ *Simplify.*

1) $\dfrac{3}{2i} =$

_____

2) $\dfrac{8}{-3i} =$

_____

3) $\dfrac{-9}{2i} =$

_____

4) $\dfrac{2-3i}{-5i} =$

_____

5) $\dfrac{4-5i}{-2i} =$

_____

6) $\dfrac{8+3i}{2i} =$

_____

7) $\dfrac{6i}{-2+3i} =$

_____

8) $\dfrac{9i}{-2-6i} =$

_____

9) $\dfrac{-8-3i}{-2+3i} =$

_____

10) $\dfrac{-6-3i}{2+5i} =$

_____

11) $\dfrac{-5-3i}{2-6i} =$

_____

12) $\dfrac{-7+2i}{2-5i} =$

_____

13) $\dfrac{-8-6i}{2-5i} =$

_____

14) $\dfrac{-6-8i}{3+5i} =$

_____

15) $\dfrac{-8+4i}{9-2i} =$

_____

16) $\dfrac{-10-3i}{-9+6i} =$

_____

Rationalizing Imaginary Denominators - Answers

✎ *Simplify.*

1) $\dfrac{3}{2i} =$

$\dfrac{3i}{2}$

2) $\dfrac{8}{-3i} =$

$\dfrac{8i}{3}$

3) $\dfrac{-9}{2i} =$

$-\dfrac{9i}{2}$

4) $\dfrac{2-3i}{-5i} =$

$\dfrac{3}{5} + \dfrac{2}{5}i$

5) $\dfrac{4-5i}{-2i} =$

$\dfrac{5}{2} + 2i$

6) $\dfrac{8+3i}{2i} =$

$-\dfrac{3}{2} + 4i$

7) $\dfrac{6i}{-2+3i} =$

$\dfrac{18}{13} - \dfrac{12}{13}i$

8) $\dfrac{9i}{-2-6i} =$

$-\dfrac{27}{20} - \dfrac{9}{20}i$

9) $\dfrac{-8-3i}{-2+3i} =$

$\dfrac{7}{13} + \dfrac{30}{13}i$

10) $\dfrac{-6-3i}{2+5i} =$

$-\dfrac{27}{29} + \dfrac{24}{29}i$

11) $\dfrac{-5-3i}{2-6i} =$

$\dfrac{1}{5} - \dfrac{9}{10}i$

12) $\dfrac{-7+2i}{2-5i} =$

$-\dfrac{24}{29} - \dfrac{31}{29}i$

13) $\dfrac{-8-6i}{2-5i} =$

$\dfrac{14}{29} - \dfrac{52}{29}i$

14) $\dfrac{-6-8i}{3+5i} =$

$-\dfrac{29}{17} + \dfrac{3}{17}i$

15) $\dfrac{-8+4i}{9-2i} =$

$-\dfrac{16}{17} + \dfrac{4}{17}i$

16) $\dfrac{-10-3i}{-9+6i} =$

$\dfrac{8}{13} + \dfrac{29}{39}i$

Simplifying Radical Expressions

✎ **Simplify.**

1) $\sqrt{256y} = $ _____

2) $\sqrt{900} = $ _____

3) $\sqrt{144a^2b} = $ _____

4) $\sqrt{36 \times 9} = $ _____

5) $\sqrt{49xy} = $ _____

6) $\sqrt{32x^4} = $ _____

7) $\sqrt{18b^2c^2} = $ _____

8) $\sqrt{100a} = $ _____

9) $\sqrt{490 \times 10} = $ _____

10) $\sqrt{a^2b^2c^6} = $ _____

11) $\sqrt{15x^4z^4} = $ _____

12) $\sqrt{25ab^2c^2} = $ _____

13) $\sqrt{100 \times 9y} = $ _____

14) $\sqrt{81b^2c^5} = $ _____

✎ **Write each radical in exponential form.**

15) $\sqrt{x} = $ _____

16) $\sqrt[3]{a^2} = $ _____

17) $\sqrt{(ab)^3} = $ _____

18) $\sqrt[5]{x^6} = $ _____

19) $\sqrt[3]{a^9} = $ _____

20) $\sqrt[7]{x^{28}} = $ _____

21) $\sqrt{x^3} = $ _____

22) $\left(\sqrt[3]{3x}\right)^5 = $ _____

23) $\sqrt[5]{2a} = $ _____

24) $\left(\frac{1}{\sqrt[3]{x}}\right)^{-1} = $ _____

25) $\frac{1}{\left(\sqrt[2]{x}\right)^4} = $ _____

26) $\sqrt[3]{5xy} = $ _____

Simplifying Radical Expressions - Answers

✎ **Simplify.**

1) $\sqrt{256y} = 16\sqrt{y}$

2) $\sqrt{900} = 30$

3) $\sqrt{144a^2b} = 12a\sqrt{b}$

4) $\sqrt{36 \times 9} = 18$

5) $\sqrt{49xy} = 7\sqrt{xy}$

6) $\sqrt{32x^4} = 4x^2\sqrt{2}$

7) $\sqrt{18b^2c^2} = 3bc\sqrt{2}$

8) $\sqrt{100a} = 10\sqrt{a}$

9) $\sqrt{490 \times 10} = 70$

10) $\sqrt{a^2b^2c^6} = abc^3$

11) $\sqrt{15x^4z^4} = x^2z^2\sqrt{15}$

12) $\sqrt{25ab^2c^2} = 5bc\sqrt{a}$

13) $\sqrt{100 \times 9y} = 30\sqrt{y}$

14) $\sqrt{81b^2c^5} = 9bc^2\sqrt{c}$

✎ **Write each radical in exponential form.**

15) $\sqrt{x} = x^{\frac{1}{2}}$

16) $\sqrt[3]{a^2} = a^{\frac{2}{3}}$

17) $\sqrt{(ab)^3} = (ab)^{\frac{3}{2}}$

18) $\sqrt[5]{x^6} = x^{\frac{6}{5}}$

19) $\sqrt[3]{a^9} = a^3$

20) $\sqrt[7]{x^{28}} = x^4$

21) $\sqrt{x^3} = x^{\frac{3}{2}}$

22) $\left(\sqrt[3]{3x}\right)^5 = (3x)^{\frac{5}{3}}$

23) $\sqrt[5]{2a} = (2a)^{\frac{1}{5}}$

24) $\left(\frac{1}{\sqrt[3]{x}}\right)^{-1} = x^{\frac{1}{3}}$

25) $\frac{1}{\left(\sqrt[2]{x}\right)^4} = x^{-2}$

26) $\sqrt[3]{5xy} = (5xy)^{\frac{1}{3}}$

Adding and Subtracting Radical Expressions

✎ *Simplify.*

1) $3\sqrt{5} + 2\sqrt{5} =$ _____

2) $6\sqrt{3} + 4\sqrt{27} =$ _____

3) $5\sqrt{2} + 10\sqrt{18} =$ _____

4) $7\sqrt{2} - 5\sqrt{8} =$ _____

5) $2\sqrt{3} + 3\sqrt{3} =$ _____

6) $9\sqrt{6} + \sqrt{54} =$ _____

7) $-4\sqrt{2} - 8\sqrt{8} =$ _____

8) $5\sqrt{7} - \sqrt{63} =$ _____

9) $-2\sqrt{18} - 5\sqrt{2} =$ _____

10) $\sqrt{28} - 5\sqrt{7} =$ _____

11) $10\sqrt{2} + \sqrt{8} =$ _____

12) $9\sqrt{3} - 2\sqrt{12} =$ _____

✎ *Find the value of* x *in each equation.*

13) $\sqrt{x} + 3\sqrt{5} = 4\sqrt{5}$, $x =$ _____

14) $x\sqrt{3} + 2\sqrt{3} = -3\sqrt{3}$, $x =$ _____

15) $\sqrt{28} + x = 4\sqrt{7}$, $x =$ _____

16) $\sqrt{12} + 3\sqrt{27} = x$, $x =$ _____

17) $2\sqrt{x} + 3\sqrt{12} = 8\sqrt{3}$, $x =$ _____

18) $x - \sqrt{2} = 5\sqrt{2}$, $x =$ _____

Adding and Subtracting Radical Expressions - Answers

✎ *Simplify.*

1) $3\sqrt{5} + 2\sqrt{5} = 5\sqrt{5}$

2) $6\sqrt{3} + 4\sqrt{27} = 18\sqrt{3}$

3) $5\sqrt{2} + 10\sqrt{18} = 35\sqrt{2}$

4) $7\sqrt{2} - 5\sqrt{8} = -3\sqrt{2}$

5) $2\sqrt{3} + 3\sqrt{3} = 5\sqrt{3}$

6) $9\sqrt{6} + \sqrt{54} = 12\sqrt{6}$

7) $-4\sqrt{2} - 8\sqrt{8} = -20\sqrt{2}$

8) $5\sqrt{7} - \sqrt{63} = 2\sqrt{7}$

9) $-2\sqrt{18} - 5\sqrt{2} = -11\sqrt{2}$

10) $\sqrt{28} - 5\sqrt{7} = -3\sqrt{7}$

11) $10\sqrt{2} + \sqrt{8} = 12\sqrt{2}$

12) $9\sqrt{3} - 2\sqrt{12} = 5\sqrt{3}$

✎ *Find the value of x in each equation.*

13) $\sqrt{x} + 3\sqrt{5} = 4\sqrt{5}$, $x = 5$

14) $x\sqrt{3} + 2\sqrt{3} = -3\sqrt{3}$, $x = -5$

15) $\sqrt{28} + x = 4\sqrt{7}$, $x = 2\sqrt{7}$

16) $\sqrt{12} + 3\sqrt{27} = x$, $x = 11\sqrt{3}$

17) $2\sqrt{x} + 3\sqrt{12} = 8\sqrt{3}$, $x = 3$

18) $x - \sqrt{2} = 5\sqrt{2}$, $x = 6\sqrt{2}$

Multiplying Radical Expressions

✍ *Evaluate.*

1) $\sqrt{5} \times \sqrt{3} =$ _____

2) $\sqrt{6} \times \sqrt{8} =$ _____

3) $3\sqrt{5} \times \sqrt{9} =$ _____

4) $2\sqrt{3} \times 3\sqrt{7} =$ _____

5) $5\sqrt{2} \times \sqrt{4} =$ _____

6) $\sqrt{10} \times 4\sqrt{6} =$ _____

7) $9\sqrt{8} \times \sqrt{5x} =$ _____

8) $\sqrt{3x} \times 2\sqrt{6x} =$ _____

9) $15\sqrt{3} \times \sqrt{5} =$ _____

10) $\sqrt{25k^2} \times \sqrt{36} =$ _____

11) $4\sqrt{6x^3} \times 7\sqrt{6x} =$ _____

12) $-6\sqrt{2} \times \sqrt{8} =$ _____

13) $-7\sqrt{25x} \times \sqrt{2x} =$ _____

14) $4\sqrt{36} \times 2\sqrt{81} =$ _____

15) $-5\sqrt{16} \times (-\sqrt{25}) =$ _____

16) $-3\sqrt{2x^3} \times (-\sqrt{x}) =$ _____

17) $\sqrt{5x} \times \sqrt{5x} =$ _____

18) $5\sqrt{2} \times \sqrt{6x} =$ _____

19) $-6\sqrt{7a^3} \times \sqrt{7a} =$ _____

20) $-\sqrt{9b} \times (-\sqrt{81b}) =$ _____

21) $4\sqrt{5} \times 3\sqrt{9} =$ _____

22) $\sqrt{y^5} \times 2\sqrt{y} =$ _____

23) $-4\sqrt{3} \times 7\sqrt{5} =$ _____

24) $2\sqrt{x^3y} \times 3\sqrt{xy^3} =$ _____

Multiplying Radical Expressions - Answers

✎ *Evaluate.*

1) $\sqrt{5} \times \sqrt{3} = \sqrt{15}$

2) $\sqrt{6} \times \sqrt{8} = \sqrt{48} = 4\sqrt{3}$

3) $3\sqrt{5} \times \sqrt{9} = 9\sqrt{5}$

4) $2\sqrt{3} \times 3\sqrt{7} = 6\sqrt{21}$

5) $5\sqrt{2} \times \sqrt{4} = 10\sqrt{2}$

6) $\sqrt{10} \times 4\sqrt{6} = 4\sqrt{60} = 8\sqrt{15}$

7) $9\sqrt{8} \times \sqrt{5x} = 18\sqrt{10x}$

8) $\sqrt{3x} \times 2\sqrt{6x} = 6x\sqrt{2}$

9) $15\sqrt{3} \times \sqrt{5} = 15\sqrt{15}$

10) $\sqrt{25k^2} \times \sqrt{36} = 30k$

11) $4\sqrt{6x^3} \times 7\sqrt{6x} = 168x^2$

12) $-6\sqrt{2} \times \sqrt{8} = -24$

13) $-7\sqrt{25x} \times \sqrt{2x} = -35x\sqrt{2}$

14) $4\sqrt{36} \times 2\sqrt{81} = 432$

15) $-5\sqrt{16} \times (-\sqrt{25}) = 100$

16) $-3\sqrt{2x^3} \times (-\sqrt{x}) = 3x^2\sqrt{2}$

17) $\sqrt{5x} \times \sqrt{5x} = 5x$

18) $5\sqrt{2} \times \sqrt{6x} = 10\sqrt{3}$

19) $-6\sqrt{7a^3} \times \sqrt{7a} = -42a^2$

20) $-\sqrt{9b} \times (-\sqrt{81b}) = 27b$

21) $4\sqrt{5} \times 3\sqrt{9} = 36\sqrt{5}$

22) $\sqrt{y^5} \times 2\sqrt{y} = 2y^3$

23) $-4\sqrt{3} \times 7\sqrt{5} = -28\sqrt{15}$

24) $2\sqrt{x^3y} \times 3\sqrt{xy^3} = 6x^2y^2$

Simplifying Radical Expressions Involving Fractions

✎ **Simplify.**

1) $\dfrac{1}{\sqrt{3}-6} =$ _____

2) $\dfrac{5}{\sqrt{2}+7} =$ _____

3) $\dfrac{\sqrt{3}}{1-\sqrt{6}} =$ _____

4) $\dfrac{2}{\sqrt{3}+5} =$ _____

5) $\dfrac{7}{\sqrt{7}-3} =$ _____

6) $\dfrac{\sqrt{5m}}{\sqrt{m^3}} =$ _____

7) $\dfrac{3\sqrt{3}}{\sqrt{x}} =$ _____

8) $\dfrac{\sqrt{8}+\sqrt{7}}{\sqrt{5}-\sqrt{2}} =$ _____

9) $\dfrac{\sqrt{6}-6}{\sqrt{6}-5} =$ _____

10) $\dfrac{2}{\sqrt{3}+5} =$ _____

11) $\dfrac{5}{3\sqrt{x}} =$ _____

12) $\dfrac{7-\sqrt{2}}{\sqrt{2}+\sqrt{6}} =$ _____

13) $\dfrac{\sqrt{y}}{\sqrt{y}+\sqrt{5}} =$ _____

14) $\dfrac{2\sqrt{6}+9}{\sqrt{2}+1} =$ _____

15) $\dfrac{7}{14\sqrt{5}} =$ _____

16) $\dfrac{10\sqrt{2}}{5\sqrt{m}} =$ _____

17) $\dfrac{\sqrt{64a^6b^5}}{\sqrt{3xy}} -$ _____

18) $\dfrac{\sqrt{25xy}}{\sqrt{x^2y^3}} =$ _____

19) $\dfrac{3\sqrt{5}-6}{\sqrt{7}-\sqrt{3}} =$ _____

20) $\dfrac{1-\sqrt{5}}{1+\sqrt{3}} =$ _____

implifying Radical Expressions Involving Fractions - Answers

✎ *Simplify.*

1) $\frac{1}{\sqrt{3}-6} = -\frac{\sqrt{3}+6}{33}$

2) $\frac{5}{\sqrt{2}+7} = -\frac{5(\sqrt{2}-7)}{47}$

3) $\frac{\sqrt{3}}{1-\sqrt{6}} = -\frac{\sqrt{3}+3\sqrt{2}}{5}$

4) $\frac{2}{\sqrt{3}+5} = -\frac{\sqrt{3}-5}{11}$

5) $\frac{7}{\sqrt{7}-3} = -\frac{7(\sqrt{7}+3)}{2}$

6) $\frac{\sqrt{5m}}{\sqrt{m^3}} = \frac{\sqrt{5}}{m}$

7) $\frac{3\sqrt{3}}{\sqrt{x}} = \frac{3\sqrt{3x}}{x}$

8) $\frac{\sqrt{8}+\sqrt{7}}{\sqrt{5}-\sqrt{2}} = \frac{2\sqrt{10}+4+\sqrt{35}+\sqrt{14}}{3}$

9) $\frac{\sqrt{6}-6}{\sqrt{6}-5} = \frac{24+\sqrt{6}}{19}$

10) $\frac{2}{\sqrt{3}+5} = -\frac{\sqrt{3}-5}{11}$

11) $\frac{5}{3\sqrt{x}} = \frac{5\sqrt{x}}{3x}$

12) $\frac{7-\sqrt{2}}{\sqrt{2}+\sqrt{6}} = \frac{7\sqrt{6}+2-7\sqrt{2}+2\sqrt{3}}{4}$

13) $\frac{\sqrt{y}}{\sqrt{y}+\sqrt{5}} = \frac{\sqrt{y}(\sqrt{y}-5)}{y-5}$

14) $\frac{2\sqrt{6}+9}{\sqrt{2}+1} = 4\sqrt{3} - 2\sqrt{6} + 9\sqrt{2} - 9$

15) $\frac{7}{14\sqrt{5}} = \frac{\sqrt{5}}{10}$

16) $\frac{10\sqrt{2}}{5\sqrt{m}} = \frac{2\sqrt{2m}}{m}$

17) $\frac{\sqrt{64a^6b^5}}{\sqrt{3ab}} = \frac{8a^2b^2\sqrt{3a}}{3}$

18) $\frac{\sqrt{25xy}}{\sqrt{x^2y^3}} = \frac{5\sqrt{x}}{xy}$

19) $\frac{3\sqrt{5}-6}{\sqrt{7}-\sqrt{3}} = \frac{(3\sqrt{5}-6)(\sqrt{7}+\sqrt{3})}{4}$

20) $\frac{1-\sqrt{5}}{1+\sqrt{3}} = -\frac{(1-\sqrt{5})(1-\sqrt{3})}{2}$

Radical Equations

Solve for x *in each equation.*

1) $\sqrt{x} + 2 = 9, x =$ _____

2) $3 + \sqrt{x} = 12, x =$ _____

3) $\sqrt{x} + 5 = 30, x =$ _____

4) $\sqrt{x} - 9 = 27, x =$ _____

5) $10 = \sqrt{x + 1}, x =$ _____

6) $\sqrt{x + 4} = 3, x =$ _____

7) $\sqrt{x + 9} = 2, x =$ _____

8) $\sqrt{x - 3} = 5, x =$ _____

9) $\sqrt{x - 10} = 6, x =$ _____

10) $2\sqrt{x} = 6, x =$ _____

11) $\sqrt{3x} = 9, x =$ _____

12) $5\sqrt{x} = 10, x =$ _____

13) $\sqrt{2x} = \sqrt{x + 4}, x =$ _____

14) $\sqrt{x + 5} = \sqrt{6 - x}, x =$ _____

15) $12 = \sqrt{x + 9}, x =$ _____

16) $4\sqrt{2x} = 16, x =$ _____

17) $\sqrt{5x - 7} = \sqrt{x + 1}, x =$ _____

18) $\sqrt{8x} = \sqrt{x + 14}, x =$ _____

19) $\sqrt{5x + 2} = \sqrt{2x + 6}, x =$ _____

20) $\sqrt{x} = \sqrt{7x - 6}, x =$ _____

21) $\sqrt{3x + 3} = \sqrt{6x}, x =$ _____

22) $\sqrt{x + 12} = \sqrt{7x - 6}, x =$ ____

Radical Equations - Answers

✎ **Solve for** x **each equation.**

1) $\sqrt{x} + 2 = 9, x = 49$

2) $3 + \sqrt{x} = 12, x = 81$

3) $\sqrt{x} + 5 = 30, x = 625$

4) $\sqrt{x} - 9 = 27, x = 1,296$

5) $10 = \sqrt{x + 1}, x = 99$

6) $\sqrt{x + 4} = 3, x = 5$

7) $\sqrt{x + 9} = 2, x = -5$

8) $\sqrt{x - 3} = 5, x = 28$

9) $\sqrt{x - 10} = 6, x = 46$

10) $2\sqrt{x} = 6, x = 9$

11) $\sqrt{3x} = 9, x = 27$

12) $5\sqrt{x} = 10, x = 4$

13) $\sqrt{2x} = \sqrt{x + 4}, x = 4$

14) $\sqrt{x + 5} = \sqrt{6 - x}, x = \frac{1}{2}$

15) $12 = \sqrt{x + 9}, x = 135$

16) $4\sqrt{2x} = 16, x = 8$

17) $\sqrt{5x - 7} = \sqrt{x + 1}, x = 2$

18) $\sqrt{8x} = \sqrt{x + 14}, x = 2$

19) $\sqrt{5x + 2} = \sqrt{2x + 6}, x = \frac{4}{3}$

20) $\sqrt{x} = \sqrt{7x - 6}, x = 1$

21) $\sqrt{3x + 3} = \sqrt{6x}, x = 1$

22) $\sqrt{x + 12} = \sqrt{7x - 6}, x = 3$

Domain and Range of Radical Functions

✎ *Identify the domain and range of each function.*

1) $y = \sqrt{x + 2} - 1$

2) $y = \sqrt{x + 1}$

3) $y = \sqrt{x - 4}$

4) $y = \sqrt{x - 3} + 1$

5) $y = \sqrt{x + 3} - 2$

6) $y = \sqrt{x + 1} - 3$

7) $y = \sqrt{x + 1} - 6$

8) $y = \sqrt{x - 2} - 5$

✎ *Sketch the graph of each function.*

9) $y = \sqrt{x} - 1$

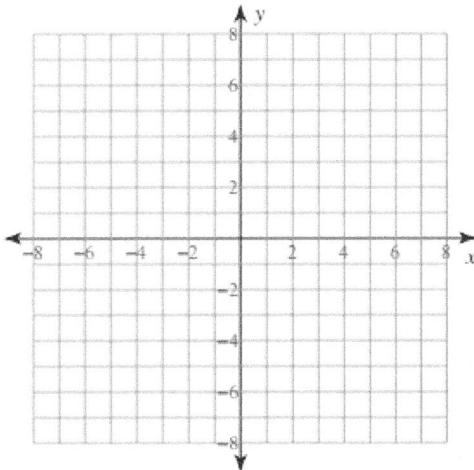

10) $y = \sqrt{x} + 2$

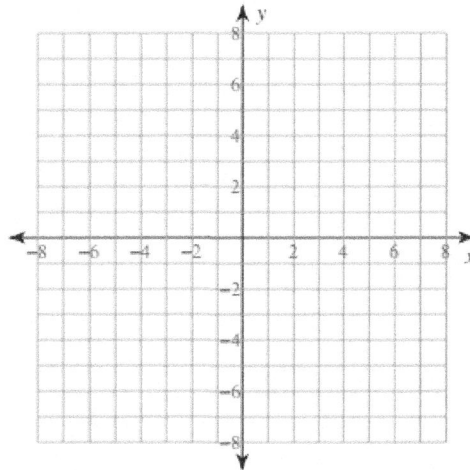

EffortlessMath.com

Domain and Range of Radical Functions - Answers

✎ *Identify the domain and range of each function.*

1) $y = \sqrt{x+2} - 1, x \geq -2, y \geq -1$

2) $y = \sqrt{x+1}, x \geq -1, y \geq 0$

3) $y = \sqrt{x-4}, x \geq 4, y \geq 0$

4) $y = \sqrt{x-3} + 1, x \geq 3, y \geq 1$

5) $y = \sqrt{x+3} - 2, x \geq -3, y \geq -2$

6) $y = \sqrt{x+1} - 3, x \geq -1, y \geq -3$

7) $y = \sqrt{x+1} - 6, x \geq -1, y \geq -6$

8) $y = \sqrt{x-2} - 5, x \geq 2, y \geq -5$

✎ *Sketch the graph of each function.*

9) $y = \sqrt{x} - 1$

10) $y = \sqrt{x} + 2$

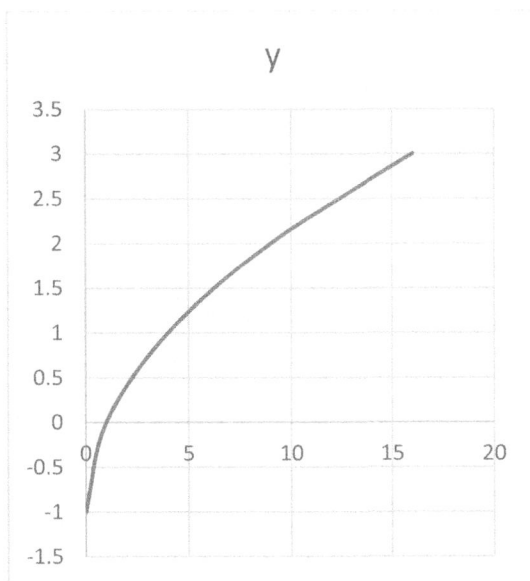

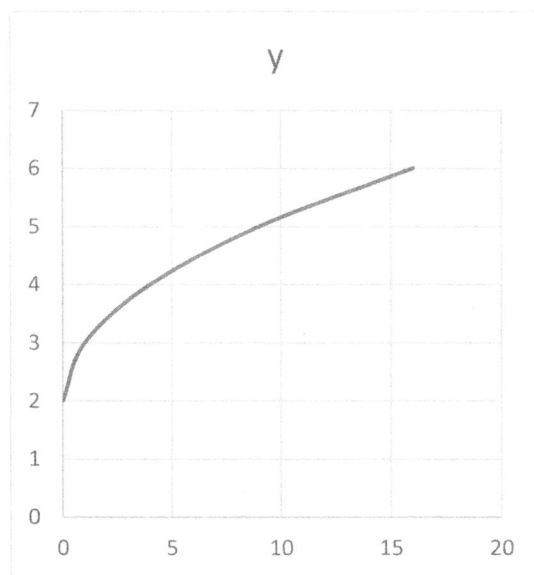

Properties Of Logarithms

✎ *Expand each logarithm.*

1) $\log_b(2 \times 9) = \underline{\hspace{2cm}}$

2) $\log_b(5 \times 7) = \underline{\hspace{2cm}}$

3) $\log_b(xy) = \underline{\hspace{2cm}}$

4) $\log_b(16 \times 2^3) = \underline{\hspace{2cm}}$

5) $\log_b(27 \times 3^2) = \underline{\hspace{2cm}}$

6) $\log_b\left(\frac{1}{9}\right) = \underline{\hspace{2cm}}$

7) $\log_a\left(\frac{3}{7}\right) = \underline{\hspace{2cm}}$

8) $\log_b\left(\frac{2}{5}\right)^3 = \underline{\hspace{2cm}}$

9) $\log_b\left(\frac{1}{4}\right)^2 = \underline{\hspace{2cm}}$

10) $\log_b\left(\frac{1}{6}\right)^2 = \underline{\hspace{2cm}}$

11) $\log_b(64 \times 4^2) = \underline{\hspace{2cm}}$

12) $3^{\log_3 6} = \underline{\hspace{2cm}}$

13) $\log_b\left(\frac{1}{5}\right)^4 = \underline{\hspace{2cm}}$

✎ *Condense each expression to a single logarithm.*

14) $\log_b 7 - \log_b 2 =$

15) $3\log_b 3 + 2\log_b 5 =$

16) $\log_b 10 - \log_b 6 =$

17) $\log_b 5 + \log_b 2 =$

18) $5\log_b 2 - 2\log_b 8 =$

19) $\log_b 2 - 5\log_b 3 =$

20) $\log_b 9 - 3\log_b\left(\frac{1}{3}\right)^2 =$

21) $\log_a 9 - 2\log_a 7 =$

22) $\log_x \frac{1}{5} + \log_x \frac{1}{3} =$

23) $\log_b 3 - \log_b 5 =$

24) $6\log_b(2^4 \times 64) =$

25) $3\log_b 9 - 6\log 4 =$

Properties of Logarithms - Answers

✍ *Expand each logarithm.*

1) $log_b(2 \times 9) = log_b 2 + 2log_b 3$

2) $log_b(5 \times 7) = log_b 5 + log_b 7$

3) $log_b(xy) = log_b x + log_b y$

4) $log_b(16 \times 2^3) = 7log_b 2$

5) $log_b(27 \times 3^2) = 5log_b 3$

6) $log_b\left(\frac{1}{9}\right) = -2log_b 3$

7) $log_a\left(\frac{3}{7}\right) = log_b 3 - log_b 7$

8) $log_b\left(\frac{2}{5}\right)^3 = 3log_b 2 - 3log_b 5$

9) $log_b\left(\frac{1}{4}\right)^2 = -4log_b 2$

10) $log_b\left(\frac{1}{6}\right)^2 = -2log_b 6$

11) $log_b(64 \times 4^2) = 5log_b 4$

12) $3^{log_3 6} = 6$

13) $log_b\left(\frac{1}{5}\right)^4 = -4log_b 5$

✍ *Condense each expression to a single logarithm.*

14) $log_b 7 - log_b 2 = log_b \frac{7}{2}$

15) $3log_b 3 + 2log_b 5 = log_b(675)$

16) $log_b 10 - log_b 6 = log_b \frac{5}{3}$

17) $log_b 5 + log_b 2 = log_b(10)$

18) $5log_b 2 - 2log_b 8 = log_b\left(\frac{2^5}{8^2}\right)$

19) $log_b 2 - 5log_b 3 = log_b\left(\frac{2}{3^5}\right)$

20) $log_b 9 - 3log_b\left(\frac{1}{3}\right)^2 = 8log_b 3$

21) $log_a 9 - 2log_a 7 = log_a\left(\frac{9}{7^2}\right)$

22) $log_x \frac{1}{5} + log_x \frac{1}{3} = log_x\left(\frac{1}{15}\right)$

23) $log_b 3 - log_b 5 = log_b \frac{3}{5}$

24) $6\, log_b(2^4 \times 64) = 60log_b 2$

25) $3log_b 9 - 6log 4 = log_b\left(\frac{9^3}{4^6}\right)$

Evaluating Logarithm

✎ **Evaluate each logarithm.**

1) $2\log_9(9) =$ _____

2) $3\log_2(8) =$ _____

3) $2\log_5(125) =$ _____

4) $\log_{100}(1) =$ _____

5) $\log_{10}(100) =$ _____

6) $3\log_4(16) =$ _____

7) $\frac{1}{2}\log_3(81) =$ _____

8) $\log_7(343) =$ _____

9) $\log_5(625) =$ _____

10) $\log_3(729) =$ _____

11) $\log_2(256) =$ _____

12) $\log_3(243) =$ _____

13) $\frac{1}{2}\log_9(81) =$ _____

14) $\log_8(512) =$ _____

15) $\log_4(1024) =$ _____

16) $\log_9(729) =$ _____

✎ **Circle the points which are on the graph of the given logarithmic functions.**

17) $y = \log_2(x - 6) + 1$ $(2, 1)$, $(4, 1)$, $(8, 2)$

18) $y = \log_4(2x)$ $(2, 1)$, $(8, 2)$, $(0, 3)$

19) $y = \log_3(2x - 1) + 4$ $(1, 4)$, $(2, 5)$, $(0, 1)$

20) $y = \log_5(x - 1)$ $(4, 1)$, $(3, 1)$, $(6, 1)$

21) $y = 2\log_3(x + 1)$ $(3, 1)$, $(2, 2)$, $(4, 2)$

22) $y = -1 + 3\log_4(2x - 2)$ $(2, 3)$, $(3, 2)$, $(4, 2)$

23) $y = -1 + 5\log_6(3x)$ $(4, 2)$, $(2, 4)$, $(1, 4)$

24) $y = -3 + \log_5(2x + 1)$ $(0, -2)$, $(1, -2)$, $(2, -2)$

Evaluating Logarithm - Answers

✎ *Evaluate each logarithm.*

1) $2\log_9(9) = 2$

2) $3\log_2(8) = 9$

3) $2\log_5(125) = 6$

4) $\log_{100}(1) = 0$

5) $\log_{10}(100) = 2$

6) $3\log_4(16) = 6$

7) $\frac{1}{2}\log_3(81) = 2$

8) $\log_7(343) = 3$

9) $\log_5(625) = 4$

10) $\log_3(729) = 6$

11) $\log_2(256) = 8$

12) $\log_3(243) = 5$

13) $\frac{1}{2}\log_9(81) = 1$

14) $\log_8(512) = 3$

15) $\log_4(1024) = 5$

16) $\log_9(729) = 3$

✎ *Circle the points which are on the graph of the given logarithmic functions.*

17) $y = \log_2(x - 6) + 1$ $(2, 1),$ $(4, 1),$ $\boxed{(8, 2)}$

18) $y = \log_4(2x)$ $\boxed{(2, 1),}$ $(8, 2),$ $(0, 3)$

19) $y = \log_3(2x - 1) + 4$ $(1, 4),$ $\boxed{(2, 5),}$ $(0, 1)$

20) $y = \log_5(x - 1)$ $(4, 1),$ $(3, 1),$ $\boxed{(6, 1)}$

21) $y = 2\log_3(x + 1)$ $(3, 1),$ $\boxed{(2, 2),}$ $(4, 2)$

22) $y = -1 + 3\log_4(2x - 2)$ $(2, 3),$ $(4, 2),$ $\boxed{(3, 2)}$

23) $y = -1 + 5\log_6(3x)$ $(4, 2),$ $\boxed{(2, 4),}$ $(1, 4)$

24) $y = -3 + \log_5(2x + 1)$ $(0, -2),$ $(1, -2),$ $\boxed{(2, -2)}$

Natural Logarithms

✎ **Solve each equation for** x.

1) $e^x = 49$, $x =$ _____

2) $e^x = 15$, $x =$ _____

3) $e^x = 3$, $x =$ _____

4) $e^x = 8$, $x =$ _____

5) $\ln x = 15$, $x =$ _____

6) $\ln x = 6$, $x =$ _____

7) $\ln(\ln x) = 9$, $x =$ _____

8) $\ln(3x - 1) = 1$, $x =$ ____

9) $\ln(4x + 6) = 1$, $x =$ ____

10) $\ln(7x - 1) = 1$, $x =$ ____

✎ **Reduce the following expressions to simplest form.**

11) $e^{\ln 4 + \ln 5} =$

12) $e^{\ln\left(\frac{9}{e}\right)} =$

13) $e^{\ln 2 + \ln 7} =$

14) $6\ln(e^5) =$

15) $e^{\ln\left(\frac{5}{e}\right)} =$

16) $\ln\left(\frac{1}{e}\right)^8 =$

17) $3\ln(e^7) =$

18) $\ln\left(\frac{1}{e}\right)^6 =$

19) $e^{\ln 2 + \ln 9} =$

✎ **If** $e^{\ln 5 + \ln x} = 35$, **then which of following is** x?

A) 5

B) 4

C) 7

D) 6

Natural Logarithms - Answers

✍ *Solve each equation for* x.

1) $e^x = 49$, $x = 2ln7$

2) $e^x = 15$, $x = ln15$

3) $e^x = 3$, $x = ln3$

4) $e^x = 8$, $x = 3ln2$

5) $ln\, x = 15$, $x = e^{15}$

6) $ln\, x = 6$, $x = e^6$

7) $ln(ln\, x) = 9$, $x = e^{e^9}$

8) $ln(3x - 1) = 1$, $x = \frac{e+1}{3}$

9) $ln(4x + 6) = 1$, $x = \frac{e-6}{4}$

10) $ln(7x - 1) = 1$, $x = \frac{e+1}{7}$

✍ *Reduce the following expressions to simplest form.*

11) $e^{ln4+ln5} = 20$

12) $e^{ln(\frac{9}{e})} = \frac{9}{e}$

13) $e^{ln2+ln7} = 14$

14) $6\, ln(e^5) = 30$

15) $e^{ln(\frac{5}{e})} = \frac{5}{e}$

16) $ln(\frac{1}{e})^8 = -8$

17) $3\, ln(e^7) = 21$

18) $ln(\frac{1}{e})^6 = -6$

19) $e^{ln2+ln9} = 18$

✍ **If** $e^{ln5+lnx} = 35$, **then which of following is** x? C

A) 5

B) 4

C) 7

D) 6

Solving Logarithmic Equations

✎ *Find the value of the variables in each equation.*

1) $log_3 8x = 3$, $x = $ ____

2) $log_4 2x = 5$, $x = $ ____

3) $log_4 5x = 0$, $x = $ ____

4) $log 4x = log 5$, $x = $ ____

5) $log 3x = log(x - 2)$, $x = $ ____

6) $log 6 + log x = 2$, $x = $ ____

7) $log x + log 5 = 0$, $x = $ ____

8) $log(6x + 1) = log(x + 3)$, $x = $ ____

9) $log(5x - 2) = log(6x - 3)$, $x = $ ____

10) $log(x - 1) = log(2x)$, $x = $ ____

11) $log 4x = log 3x - 1$, $x = $ ____

12) $log(3x - 4) = log(2x + 5)$, $x = $ ____

13) $log 8x = log 6$, $x = $ ____

14) $log 7x = log(x + 6)$, $x = $ ____

15) $log 2x + log 3 = 0$, $x = $ ____

Solving Logarithmic Equations - Answers

✎ *Find the value of the variables in each equation.*

1) $\log_3 8x = 3$, $x = \dfrac{27}{8}$

2) $\log_4 2x = 5$, $x = 512$

3) $\log_4 5x = 0$, $x = \dfrac{1}{5}$

4) $\log 4x = \log 5$, $x = \dfrac{5}{4}$

5) $\log 3x = \log(x - 2)$, $x = $ No solution for $x \in R$

6) $\log 6 + \log x = 2$, $x = \dfrac{50}{3}$

7) $\log x + \log 5 = 0$, $x = \dfrac{1}{5}$

8) $\log(6x + 1) = \log(x + 3)$, $x = \dfrac{2}{5}$

9) $\log(5x - 2) = \log(6x - 3)$, $x = 1$

10) $\log(x - 1) = \log(2x)$, $x = $ No solution for $x \in R$

11) $\log 4x = \log 3x - 1$, $x = $ No solution for $x \in R$

12) $\log(3x - 4) = \log(2x + 5)$, $x = 9$

13) $\log 8x = \log 6$, $x = \dfrac{3}{4}$

14) $\log 7x = \log(x + 6)$, $x = 1$

15) $\log 2x + \log 3 = 0$, $x = \dfrac{1}{6}$

Arc Length and Sector Area

✎ *Find the length of each arc. Round your answers to the nearest hundredth.*

1) $r = 4$ cm, $\theta = 28°$ → arc = ___

6) $r = 3$ in, $\theta = 20°$ → arc = ____

2) $r = 6$ ft, $\theta = 30°$ → arc = ____

7) $r = 7$ cm, $\theta = 66°$ → arc = ___

3) $r = 8$ ft, $\theta = 40°$ → arc = ____

8) $r = 6$ ft, $\theta = 80°$ → arc = ____

4) $r = 12$ cm, $\theta = 34°$ → arc = __

9) $r = 11$ ft, $\theta = 68°$ → arc = ___

5) $r = 10$ in, $\theta = 70°$ → arc = __

10) $r = 5$ in, $\theta = 42°$ → arc = ___

✎ *Find area of each sector. Round your answers to the nearest tenth.*

11)

12)

15*cm*

18*ft*

Area

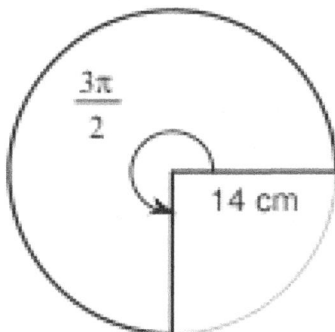

of the sector:

_____

Area of the

_____

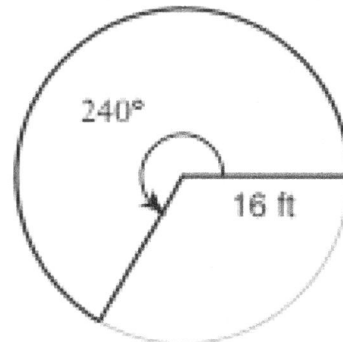

sector:

EffortlessMath.com

bit.ly/3WarzHa

Find more at

Arc Length and Sector Area - Answers

✍ *Find the length of each arc. Round your answers to the nearest hundredth.*

1) r = 4 cm, θ = 28° → arc = *1.95 cm*

2) r = 6 ft, θ = 30° → arc = *3.14 ft*

3) r = 8 ft, θ = 40° → arc = *5.58 ft*

4) r = 12 cm, θ = 34° → arc = *7.12 cm*

5) r = 10 in, θ = 70° → arc = *12.21 in*

6) r = 3 in, θ = 20° → arc = *10.05 in*

7) r = 7 cm, θ = 66° → arc = *8.06 cm*

8) r = 6 ft, θ = 80° → arc = *8.37 ft*

9) r = 11 ft, θ = 68° → arc = *13.05 ft*

10) r = 5 in, θ = 42° → arc = *3.66 in*

✍ *Find area of each sector. Round your answers to the nearest tenth.*

11)

$\frac{3\pi}{2}$ 15cm

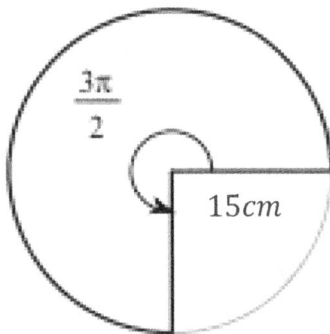

Area of a sector: 678.2 ft²

12)

240° 18ft

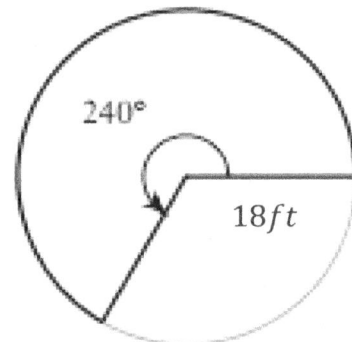

Area of a sector: 529. cm²

Equation of a Circle

✍ *Write the standard form equation of each circle.*

1) $x^2 + y^2 - 4x + 2y - 4 = 0 \rightarrow$ _____

2) $x^2 + y^2 - 8x + 6y - 11 = 0 \rightarrow$ _____

3) $x^2 + y^2 - 10x - 12y + 12 = 0 \rightarrow$ _____

4) $x^2 + y^2 + 12x - 6y - 19 = 0 \rightarrow$ _____

5) $x^2 + y^2 - 6x + 8y + 24 = 0 \rightarrow$ _____

6) $x^2 + y^2 - 14x - 8y - 16 = 0 \rightarrow$ _____

7) $x^2 + y^2 + 10x + 2y - 10 = 0 \rightarrow$ _____

8) $x^2 + y^2 - 4x + 10y - 7 = 0 \rightarrow$ _____

9) $x^2 + y^2 + 8x - 2y - 8 = 0 \rightarrow$ _____

10) $x^2 + y^2 + 4x - 4y - 1 = 0 \rightarrow$ _____

11) **Center:**$(-6,5)$, **Radius:** $4 \rightarrow$ _____

12) **Center:**$(7, -3)$, **Radius:** $3 \rightarrow$ _____

13) **Center:**$(-6,5)$, **Area:** $25\pi \rightarrow$ _____

14) **Center:**$(-6,2)$, **Area:** $49\pi \rightarrow$ _____

15) **Center:**$(-1,-5)$, **Circumference:** $2\sqrt{3}\pi \rightarrow$ _____

Equation of a Circle - Answers

✍ *Write the standard form equation of each circle.*

1) $x^2 + y^2 - 4x + 2y - 4 = 0 \rightarrow (x - 2)^2 + (y - (-1))^2 = 3^2$

2) $x^2 + y^2 - 8x + 6y - 11 = 0 \rightarrow (x - 4)^2 + (y - (-3))^2 = 6^2$

3) $x^2 + y^2 - 10x - 12y + 12 = 0 \rightarrow (x - 5)^2 + (y - 6)^2 = 7^2$

4) $x^2 + y^2 + 12x - 6y - 19 = 0 \rightarrow (x - (-6))^2 + (y - 3)^2 = 8^2$

5) $x^2 + y^2 - 6x + 8y + 24 = 0 \rightarrow (x - 3)^2 + (y - (-4))^2 = 1^2$

6) $x^2 + y^2 - 14x - 8y - 16 = 0 \rightarrow (x - 7)^2 + (y - 4)^2 = 9^2$

7) $x^2 + y^2 + 10x + 2y - 10 = 0 \rightarrow (x - (-5))^2 + (y - (-1))^2 = 6^2$

8) $x^2 + y^2 - 4x + 10y - 7 = 0 \rightarrow (x - 2)^2 + (y - (-5))^2 = 6^2$

9) $x^2 + y^2 + 8x - 2y - 8 = 0 \rightarrow (x - (-4))^2 + (y - 1)^2 = 5^2$

10) $x^2 + y^2 + 4x - 4y - 1 = 0 \rightarrow (x - (-2))^2 + (y - 2)^2 = 3^2$

11) **Center:** $(-6, 5)$, **Radius:** $4 \rightarrow (x - (-6))^2 + (y - 5)^2 = 4^2$

12) **Center:** $(7, -3)$, **Radius:** $3 \rightarrow (x - 7)^2 + (y - (-3))^2 = 3^2$

13) **Center:** $(-6, 5)$, **Area:** $25\pi \rightarrow (x - (-6))^2 + (y - 5)^2 = 5^2$

14) **Center:** $(-6, 2)$, **Area:** $49\pi \rightarrow (x - (-6))^2 + (y - 2)^2 = 7^2$

15) **Center:** $(-1, -5)$, **Circumference:** $2\sqrt{3}\pi \rightarrow (x - (-1))^2 + (y - (-5))^2 = 3$

Finding the Center and the Radius of Circles

✎ *Identify the center and radius of each.*

1) $(x + 1)^2 + (y - 2)^2 = 5 \rightarrow$ *Center:* (__ , __), *Radius:* ____

2) $(x - 5)^2 + (y + 10)^2 = 4 \rightarrow$ *Center:* (__ , __), *Radius:* ____

3) $x^2 + (y - 3)^2 = 8 \rightarrow$ *Center:* (__ , __), *Radius:* ____

4) $(x - 1)^2 + y^2 = 9 \rightarrow$ *Center:* (__ , __), *Radius:* ____

5) $x^2 + y^2 = 16 \rightarrow$ *Center:* (__ , __), *Radius:* ____

6) $(x + 1)^2 + (y + 6)^2 = 10 \rightarrow$ *Center:* (__ , __), *Radius:* ____

7) $x^2 + y^2 + 4x + 4 = 9 \rightarrow$ *Center:* (__ , __), *Radius:* ____

8) $x^2 + y^2 + 10x = 4y - 4 \rightarrow$ *Center:* (__ , __), *Radius:* ____

9) $x^2 + y^2 + 8x + 2y = -1 \rightarrow$ *Center:* (__ , __), *Radius:* ____

10) $x^2 + y^2 - 2x - 2y - 7 = 0 \rightarrow$ *Center:* (__ , __), *Radius:* ____

11) $x^2 + y^2 + 6x = 6y - 9 \rightarrow$ *Center:* (__ , __), *Radius:* ____

12) $x^2 + y^2 + 12x = 2y - 1 \rightarrow$ *Center:* (__ , __), *Radius:* ____

Finding the Center and the Radius of Circles - Answers

✎ **Identify the center and radius of each.**

1) $(x + 1)^2 + (y - 2)^2 = 5 \rightarrow$ Center: $(-1, 2)$, Radius: $\sqrt{5}$

2) $(x - 5)^2 + (y + 10)^2 = 4 \rightarrow$ Center: $(5, -10)$, Radius: 2

3) $x^2 + (y - 3)^2 = 8 \rightarrow$ Center: $(0, 3)$, Radius: $2\sqrt{2}$

4) $(x - 1)^2 + y^2 = 9 \rightarrow$ Center: $(1, 0)$, Radius: 3

5) $x^2 + y^2 = 16 \rightarrow$ Center: $(0, 0)$, Radius: 4

6) $(x + 1)^2 + (y + 6)^2 = 10 \rightarrow$ Center: $(-1, -6)$, Radius: $\sqrt{10}$

7) $x^2 + y^2 + 4x + 4 = 9 \rightarrow$ Center: $(-2, 0)$, Radius: 3

8) $x^2 + y^2 + 10x = 4y - 4 \rightarrow$ Center: $(-5, 2)$, Radius: 5

9) $x^2 + y^2 + 8x + 2y = -1 \rightarrow$ Center: $(-4, -1)$, Radius: 4

10) $x^2 + y^2 - 2x - 2y - 7 = 0 \rightarrow$ Center: $(1, 1)$, Radius: 3

11) $x^2 + y^2 + 6x = 6y - 9 \rightarrow$ Center: $(-3, 3)$, Radius: 3

12) $x^2 + y^2 + 12x = 2y - 1 \rightarrow$ Center: $(-6, 1)$, Radius: 6

Simplifying Complex Fractions

✎ **Simplify each expression.**

1) $\dfrac{\frac{2}{5}}{\frac{4}{7}} = $ _____

2) $\dfrac{6}{\frac{5}{x}+\frac{2}{3x}} = $ _____

3) $\dfrac{1-\frac{2}{x-1}}{1+\frac{4}{x+1}} = $ _____

4) $\dfrac{x}{\frac{3}{4}-\frac{5}{x}} = $ _____

5) $\dfrac{\frac{4}{x-1}}{\frac{5}{x^2+3x-4}} = $ _____

6) $\dfrac{\frac{x+1}{2}}{\frac{x+5}{x-1}} = $ _____

7) $\dfrac{\frac{5}{x}-\frac{4}{x}}{15} = $ _____

8) $\dfrac{\frac{10}{3}}{\frac{9}{4}} = $ _____

9) $\dfrac{10}{\frac{2}{x}+\frac{3}{4x}} = $ _____

10) $\dfrac{x}{\frac{2}{3}-\frac{3}{x}} = $ _____

11) $\dfrac{\frac{x+8}{2}}{\frac{x^2}{2}-\frac{9}{2}} = $ _____

12) $\dfrac{\frac{x-2}{x-5}}{\frac{x-4}{x+9}} = $ _____

13) $\dfrac{\frac{x-2}{x-1}}{\frac{x+1}{x-4}} = $ _____

14) $\dfrac{\frac{1}{x-4}}{\frac{x-5}{2}} = $ _____

15) $\dfrac{1+\frac{3}{x+2}}{1-\frac{5}{x+4}} = $ _____

16) $\dfrac{\frac{2}{x+3}}{\frac{6}{x^2+5x+6}} = $ _____

Simplifying Complex Fractions - Answers

✐Simplify each expression.

1) $\dfrac{\frac{2}{5}}{\frac{4}{7}} = \dfrac{7}{10}$

2) $\dfrac{6}{\frac{5}{x}+\frac{2}{3x}} = \dfrac{18x}{17}$

3) $\dfrac{1-\frac{2}{x-1}}{1+\frac{4}{x+1}} = \dfrac{x^2-2x-3}{x^2+4x-5}$

4) $\dfrac{x}{\frac{3}{4}-\frac{5}{x}} = \dfrac{4x^2}{3x-20}$

5) $\dfrac{\frac{4}{x-1}}{\frac{5}{x^2+3x-4}} = \dfrac{4x+16}{5}$

6) $\dfrac{\frac{x+1}{2}}{\frac{x+5}{x-1}} = \dfrac{x^2-1}{2x+10}$

7) $\dfrac{\frac{5}{x}-\frac{4}{x}}{15} = \dfrac{1}{15x}$

8) $\dfrac{\frac{10}{3}}{\frac{9}{4}} = \dfrac{40}{27}$

9) $\dfrac{10}{\frac{2}{x}+\frac{3}{4x}} = \dfrac{40x}{11}$

10) $\dfrac{x}{\frac{2}{3}-\frac{3}{x}} = \dfrac{3x^2}{2x-9}$

11) $\dfrac{\frac{x+8}{2}}{\frac{x^2}{2}-\frac{9}{2}} = \dfrac{x+8}{x^2-9}$

12) $\dfrac{\frac{x-2}{x-5}}{\frac{x-4}{x+9}} = \dfrac{x^2+7x-18}{x^2-9x+20}$

13) $\dfrac{\frac{x-2}{x-1}}{\frac{x+1}{x-4}} = \dfrac{x^2-6x+8}{x^2-1}$

14) $\dfrac{\frac{1}{x-4}}{\frac{x-5}{2}} = \dfrac{2}{x^2-9x+20}$

15) $\dfrac{1+\frac{3}{x+2}}{1-\frac{5}{x+4}} = \dfrac{x^2+9x+20}{x^2+x-2}$

16) $\dfrac{\frac{2}{x+3}}{\frac{6}{x^2+5x+6}} = \dfrac{x+2}{3}$

Graphing Rational Expressions

✍ *Graph rational expressions.*

1) $f(x) = \dfrac{x^2}{5x+6}$

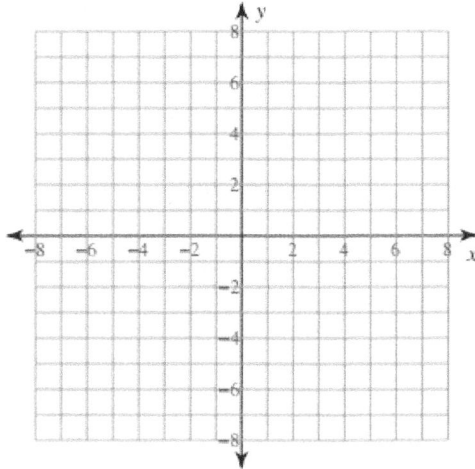

2) $f(x) = \dfrac{x^2+8x+10}{x+5}$

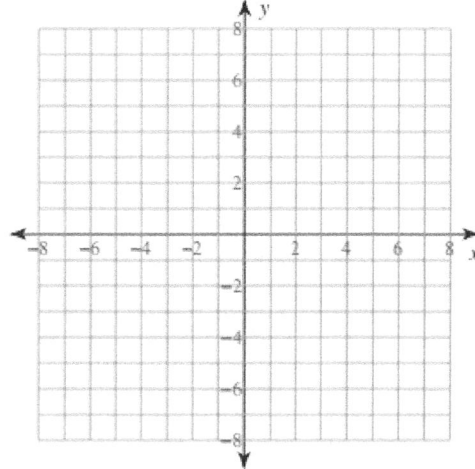

3) $f(x) = \dfrac{2x-4}{3x^2+6x+1}$

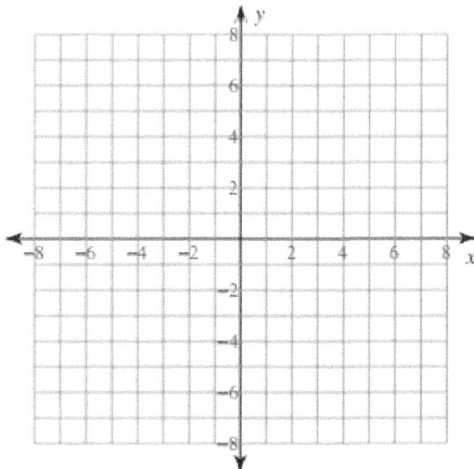

4) $f(x) = \dfrac{x^2+6x}{x+4}$

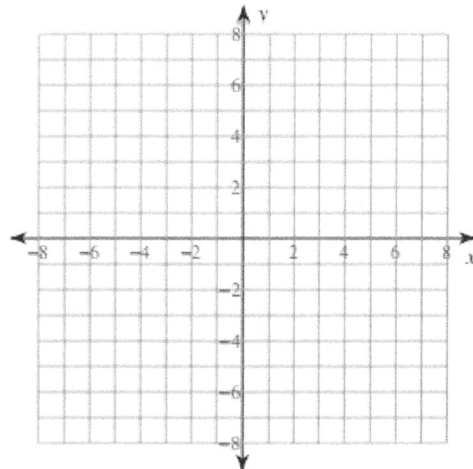

Graphing Rational Expressions - Answers

✎ *Graph rational expressions.*

1) $f(x) = \dfrac{x^2}{5x+6}$

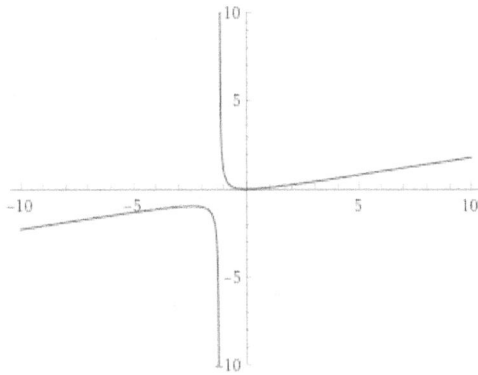

2) $f(x) = \dfrac{x^2+8x+10}{x+5}$

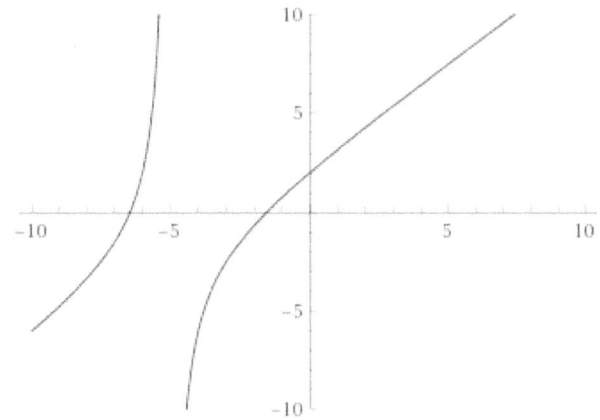

3)

$f(x) = \dfrac{2x-4}{3x^2+6x+1}$

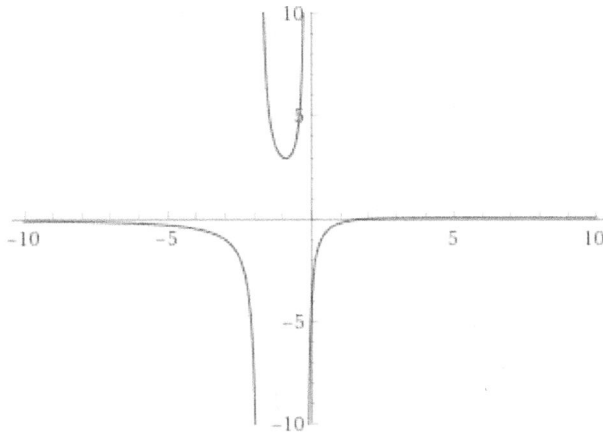

4) $f(x) = \dfrac{x^2+6x}{x+4}$

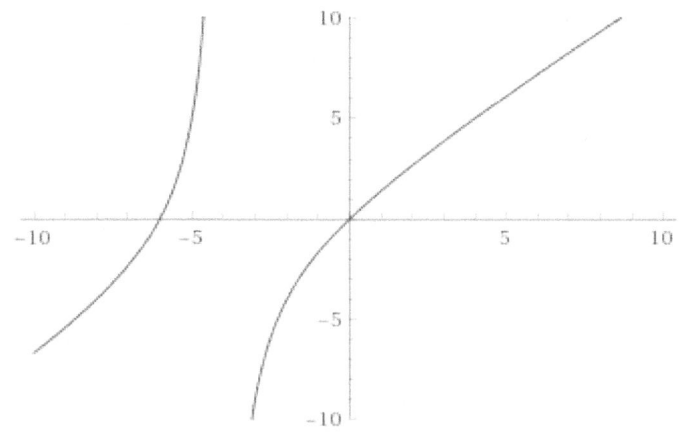

Adding and Subtracting Rational Expressions

✎ *Simplify each expression.*

1) $\dfrac{5}{x+2} + \dfrac{x-1}{x+2} =$ _____

2) $\dfrac{6}{x+5} - \dfrac{5}{x+5} =$ _____

3) $\dfrac{7}{4x+10} + \dfrac{x-5}{4x+10} =$ _____

4) $\dfrac{9x}{x+2} + \dfrac{x-1}{x+3} =$ _____

5) $\dfrac{9}{x+7} + \dfrac{x+1}{x+3} =$ _____

6) $\dfrac{x+3}{5x+2} + \dfrac{x+8}{5x+2} =$ _____

7) $\dfrac{4}{3x+2} - \dfrac{2}{x+4} =$ _____

8) $\dfrac{x+8}{x+6} + \dfrac{x-8}{x+6} =$ _____

9) $\dfrac{8}{x^2-5x+4} + \dfrac{2}{x^2-16} =$ _____

10) $\dfrac{15}{x^2-6x+5} - \dfrac{4}{x-5} =$ _____

11) $\dfrac{5}{6x-2} + \dfrac{x}{6x-2} =$ _____

12) $\dfrac{x+2}{x+3} + \dfrac{4}{x+1} =$ _____

13) $\dfrac{8}{x+3} - \dfrac{5}{x+1} =$ _____

14) $\dfrac{8x}{4x+5} - \dfrac{6x}{2x+3} =$ _____

15) $\dfrac{3x}{2x+3} - \dfrac{x}{3x+2} =$ _____

16) $\dfrac{x+6}{4x+1} - \dfrac{x-6}{4x+1} =$ _____

17) $\dfrac{9}{x+5} + \dfrac{6}{x+4} =$ _____

18) $\dfrac{6}{x+4} - \dfrac{2}{x^2-16} =$ _____

19) $\dfrac{x-4}{x^2-4} + \dfrac{x+1}{4-x^2} =$ _____

20) $\dfrac{12}{x^2-49} - \dfrac{5}{x+7} =$ _____

Adding and Subtracting Rational Expressions - Answers

✍ *Simplify each expression*.

1) $\dfrac{5}{x+2} + \dfrac{x-1}{x+2} = \dfrac{x+4}{x+2}$

2) $\dfrac{6}{x+5} - \dfrac{5}{x+5} = \dfrac{1}{x+5}$

3) $\dfrac{7}{4x+10} + \dfrac{x-5}{4x+10} = \dfrac{x+2}{4x+10}$

4) $\dfrac{9x}{x+2} + \dfrac{x-1}{x+3} = \dfrac{10x^2+28x-2}{(x+2)(x+3)}$

5) $\dfrac{9}{x+7} + \dfrac{x+1}{x+3} = \dfrac{x^2+17x+34}{(x+7)(x+3)}$

6) $\dfrac{x+3}{5x+2} + \dfrac{x+8}{5x+2} = \dfrac{2x+11}{5x+2}$

7) $\dfrac{4}{3x+2} - \dfrac{2}{x+4} = \dfrac{-2x+12}{(3x+2)(x+4)}$

8) $\dfrac{x+8}{x+6} + \dfrac{x-8}{x+6} = \dfrac{2x}{x+6}$

9) $\dfrac{8}{x^2-5x+4} + \dfrac{2}{x^2-16} = \dfrac{10x+30}{(x-4)(x+4)(x-1)}$

10) $\dfrac{15}{x^2-6x+5} - \dfrac{4}{x-5} = \dfrac{-4x+19}{(x-1)(x-5)}$

11) $\dfrac{5}{6x-2} + \dfrac{x}{6x-2} = \dfrac{x+5}{6x-2}$

12) $\dfrac{x+2}{x+3} + \dfrac{4}{x+1} = \dfrac{x^2+7x+14}{(x+3)(x+1)}$

13) $\dfrac{8}{x+3} - \dfrac{5}{x+1} = \dfrac{3x-7}{(x+3)(x+1)}$

14) $\dfrac{8x}{4x+5} - \dfrac{6x}{2x+3} = \dfrac{-8x^2-6x}{(4x+5)(2x+3)}$

15) $\dfrac{3x}{2x+3} - \dfrac{x}{3x+2} = \dfrac{7x^2+3x}{(2x+3)(3x+2)}$

16) $\dfrac{x+6}{4x+1} - \dfrac{x-6}{4x+1} = \dfrac{12}{4x+1}$

17) $\dfrac{9}{x+5} + \dfrac{6}{x+4} = \dfrac{15x+66}{(x+5)(x+4)}$

18) $\dfrac{6}{x+4} - \dfrac{2}{x^2-16} = \dfrac{6x-26}{(x+4)(x-4)}$

19) $\dfrac{x-4}{x^2-4} + \dfrac{x+1}{4-x^2} = -\dfrac{5}{(x-2)(x+2)}$

20) $\dfrac{12}{x^2-49} - \dfrac{5}{x+7} = \dfrac{-5x+47}{(x+7)(x-7)}$

Multiplying Rational Expressions

✍ *Simplify each expression.*

1) $\frac{x+1}{x+5} \times \frac{x+6}{x+1} =$ _____

2) $\frac{x+4}{x+9} \times \frac{x+9}{x+3} =$ _____

3) $\frac{x+8}{x} \times \frac{2}{x+8} =$ _____

4) $\frac{x+5}{x+1} \times \frac{x^2}{x+5} =$ _____

5) $\frac{x-3}{x+2} \times \frac{2x+4}{x+4} =$ _____

6) $\frac{x-6}{x+3} \times \frac{2x+6}{2x} =$ _____

7) $\frac{x+1}{x+5} \times \frac{5x+25}{x^2} =$ _____

8) $\frac{x+7}{x+4} \times \frac{4x+8}{x+7} =$ _____

9) $\frac{80x}{15} \times \frac{9}{10x^2} =$ _____

10) $\frac{44}{12x} \times \frac{24x}{11} =$ _____

11) $\frac{23x}{33} \times \frac{11x^2}{2} =$ _____

12) $\frac{15x^2}{20} \times \frac{12y}{3x^2} =$ _____

13) $\frac{x+9}{2x} \times \frac{4}{3x+27} =$ _____

14) $\frac{18x^2}{32} \times \frac{16}{6x} =$ _____

15) $\frac{x-6}{x+4} \times \frac{x+4}{10x-60} =$ _____

16) $\frac{4x+40}{x+10} \times \frac{x+3}{4} =$ _____

17) $\frac{1}{x+10} \times \frac{10x+60}{x+6} =$ _____

18) $\frac{x+8}{x} \times \frac{5x}{2x+16} =$ _____

19) $\frac{3x}{44} \times \frac{45x^3}{12} =$ _____

20) $\frac{x-7}{x+5} \times \frac{4x+20}{x-7} =$ _____

Multiplying Rational Expressions - Answers

✎ *simplify each expression.*

1) $\dfrac{x+1}{x+5} \times \dfrac{x+6}{x+1} = \dfrac{x+6}{x+5}$

2) $\dfrac{x+4}{x+9} \times \dfrac{x+9}{x+3} = \dfrac{x+4}{x+3}$

3) $\dfrac{x+8}{x} \times \dfrac{2}{x+8} = \dfrac{2}{x}$

4) $\dfrac{x+5}{x+1} \times \dfrac{x^2}{x+5} = \dfrac{x^2}{x+1}$

5) $\dfrac{x-3}{x+2} \times \dfrac{2x+4}{x+4} = \dfrac{2(x-3)}{x+4}$

6) $\dfrac{x-6}{x+3} \times \dfrac{2x+6}{2x} = \dfrac{x-6}{x}$

7) $\dfrac{x+1}{x+5} \times \dfrac{5x+25}{x^2} = \dfrac{5(x+1)}{x^2}$

8) $\dfrac{x+7}{x+4} \times \dfrac{4x+8}{x+7} = 2$

9) $\dfrac{80x}{15} \times \dfrac{9}{10x^2} = \dfrac{24}{5x}$

10) $\dfrac{44}{12x} \times \dfrac{24x}{11} = 8$

11) $\dfrac{23x}{33} \times \dfrac{11x^2}{2} = \dfrac{23x^3}{6}$

12) $\dfrac{15x^2}{20} \times \dfrac{12y}{3x^2} = 3y$

13) $\dfrac{x+9}{2x} \times \dfrac{4}{3x+27} = \dfrac{2}{3x}$

14) $\dfrac{18x^2}{32} \times \dfrac{16}{6x} = \dfrac{3x}{2}$

15) $\dfrac{x-6}{x+4} \times \dfrac{x+4}{10x-60} = \dfrac{1}{10}$

16) $\dfrac{4x+40}{x+10} \times \dfrac{x+3}{4} = x+3$

17) $\dfrac{1}{x+10} \times \dfrac{10x+60}{x+6} = \dfrac{10}{x+10}$

18) $\dfrac{x+8}{x} \times \dfrac{5x}{2x+16} = \dfrac{5}{2}$

19) $\dfrac{3x}{44} \times \dfrac{45x^3}{12} = \dfrac{45x^4}{176}$

20) $\dfrac{x-2}{x+5} \times \dfrac{4x+20}{x-7} = \dfrac{4(x-2)}{x-7}$

Dividing Rational Expressions

1) $\dfrac{5x}{4} \div \dfrac{5}{2} =$ _____

2) $\dfrac{8}{3x} \div \dfrac{24}{x} =$ _____

3) $\dfrac{3x}{x+4} \div \dfrac{x}{3x+12} =$ _____

4) $\dfrac{2}{5x} \div \dfrac{16}{10x} =$ _____

5) $\dfrac{36x}{5} \div \dfrac{4}{3} =$ _____

6) $\dfrac{15x^2}{6} \div \dfrac{5x}{14} =$ _____

7) $\dfrac{x-3}{x+2} \div \dfrac{x}{x+2} =$ _____

8) $\dfrac{4x}{x-8} \div \dfrac{4x}{x-2} =$ _____

9) $\dfrac{x+2}{5x^2+10x} \div \dfrac{6}{5x} =$ _____

10) $\dfrac{12x}{x-6} \div \dfrac{6}{4x-24} =$ _____

11) $\dfrac{x+4}{x+6} \div \dfrac{x^2+2x-8}{3} =$ _____

12) $\dfrac{x^2+5x+6}{x+1} \div \dfrac{x+2}{x-6} =$

13) $\dfrac{4x+16}{x+2} \div \dfrac{x^2+16x}{x+2} =$ _____

14) $\dfrac{8}{x-4} \div \dfrac{2x}{x^2-x-12} =$ _____

15) $\dfrac{7x+1}{2} \div \dfrac{70x+10}{5} =$ _____

16) $\dfrac{2x+1}{x+4} \div \dfrac{4x^2+2x}{2x+8} =$ _____

17) $\dfrac{3x-2}{x-2} \div \dfrac{9x-6}{x^2-4} =$ _____

18) $\dfrac{25x^3}{9} \div \dfrac{5x^2}{3} =$ _____

19) $\dfrac{x^2+11x+30}{x+10} \div \dfrac{x^2+3x-10}{x^2-100} =$ ___

20) $\dfrac{35x^2}{x^2-49} \div \dfrac{5x}{4x-28} =$ ___

21) $\dfrac{3x}{2} \div \dfrac{6x}{x+5} =$ _____

22) $\dfrac{2x^4}{x+6} \div \dfrac{3x^2}{x^2-36} =$ _____

Dividing Rational Expressions - Answers

1) $\dfrac{5x}{4} \div \dfrac{5}{2} = \dfrac{x}{2}$

2) $\dfrac{8}{3x} \div \dfrac{24}{x} = \dfrac{1}{9}$

3) $\dfrac{3x}{x+4} \div \dfrac{x}{3x+12} = 9$

4) $\dfrac{2}{5x} \div \dfrac{16}{10x} = \dfrac{1}{4}$

5) $\dfrac{36x}{5} \div \dfrac{4}{3} = \dfrac{27}{5}$

6) $\dfrac{15x^2}{6} \div \dfrac{5x}{14} = 7$

7) $\dfrac{x-3}{x+2} \div \dfrac{x}{x+2} = \dfrac{x-3}{x}$

8) $\dfrac{4x}{x-8} \div \dfrac{4x}{x-2} = \dfrac{x-2}{x-8}$

9) $\dfrac{x+2}{5x^2+10x} \div \dfrac{6}{5x} = \dfrac{1}{6}$

10) $\dfrac{12x}{x-6} \div \dfrac{6}{4x-24} = 8x$

11) $\dfrac{x+4}{x+6} \div \dfrac{x^2+2x-8}{3} = \dfrac{3}{(x+6)(x-2)}$

12) $\dfrac{x^2+5x+6}{x+1} \div \dfrac{x+2}{x-6} = \dfrac{(x+3)(x-6)}{x+1}$

13) $\dfrac{4x+16}{x+2} \div \dfrac{x^2+16x}{x+2} = \dfrac{4(x+4)}{x^2+16x}$

14) $\dfrac{8}{x-4} \div \dfrac{2x}{x^2-x-12} = \dfrac{4(x+3)}{x}$

15) $\dfrac{7x+1}{2} \div \dfrac{70x+10}{5} = \dfrac{1}{4}$

16) $\dfrac{2x+1}{x+4} \div \dfrac{4x^2+2x}{2x+8} = \dfrac{1}{x}$

17) $\dfrac{3x-2}{x-2} \div \dfrac{9x-6}{x^2-4} = \dfrac{x+2}{3}$

18) $\dfrac{25x^3}{9} \div \dfrac{5x^2}{3} = \dfrac{5}{3}$

19) $\dfrac{x^2+11x+30}{x+10} \div \dfrac{x^2+3x-10}{x^2-100} = \dfrac{(x+6)(x-10)}{x-2}$

20) $\dfrac{35x^2}{x^2-49} \div \dfrac{5x}{4x-28} = \dfrac{28x}{x+7}$

21) $\dfrac{3x}{2} \div \dfrac{6x}{x+5} = \dfrac{x+5}{4}$

22) $\dfrac{2x^4}{x+6} \div \dfrac{3x^2}{x^2-36} = \dfrac{2x^2(x-6)}{3}$

Rational Equations

✎ *Solve each equation.*

1) $\frac{1}{8x^2} = \frac{1}{4x^2} - \frac{1}{x} \rightarrow x =$ ____

2) $\frac{1}{9x^2} + \frac{1}{9x} = \frac{1}{x^2} \rightarrow x =$ ____

3) $\frac{32}{2x^2} + 1 = \frac{8}{x} \rightarrow x =$ ____

4) $\frac{1}{x-5} = \frac{4}{x-5} + 1 \rightarrow x =$ ____

5) $\frac{1}{x^2} + \frac{1}{x} = \frac{1}{4x^2} \rightarrow x =$ ____

6) $\frac{1}{10x^2} = \frac{1}{2x} + \frac{11}{10x^2} \rightarrow x =$ ____

7) $\frac{1}{x^2-x} + \frac{1}{x} = \frac{10}{x^2-x} \rightarrow x =$ ____

8) $\frac{x-5}{6x} = \frac{1}{5x} + 1 \rightarrow x =$ ____

9) $\frac{x-1}{x} + \frac{1}{x^2+2x} = 1 \rightarrow x =$ ____

10) $\frac{x-2}{x+3} - 1 = \frac{3}{x+2} \rightarrow x =$ ____

11) $\frac{x}{10} = \frac{6}{x-4} \rightarrow x =$ ____

12) $\frac{2}{x-2} = \frac{5}{x-1} \rightarrow x =$ ____

13) $\frac{3}{5x} = \frac{5}{9x-2} \rightarrow x =$ ____

14) $\frac{2}{2x-3} = \frac{10}{6x+1} \rightarrow x =$ ____

15) $\frac{2}{x+2} + \frac{3}{x} = \frac{x}{x+2} \rightarrow x =$ ____

16) $\frac{3}{x+1} = \frac{2}{x-3} \rightarrow x =$ ____

17) $\frac{x}{x-2} + \frac{1}{5} = \frac{x+1}{x-2} \rightarrow x =$ ____

18) $\frac{1}{x-4} + \frac{x}{x-2} = \frac{2}{x^2-6x+8} \rightarrow x =$ ____

19) $\frac{x}{x+3} = \frac{8}{x+6} \rightarrow x =$ ____

20) $\frac{x}{x-1} - \frac{2}{x} = \frac{1}{x-1} \rightarrow x =$ ____

EffortlessMath.com

Rational Equations - Answers

✎ *Solve each equation.*

1) $\frac{1}{8x^2} = \frac{1}{4x^2} - \frac{1}{x} \rightarrow x = \frac{1}{8}$

2) $\frac{1}{9x^2} + \frac{1}{9x} = \frac{1}{x^2} \rightarrow x = 8$

3) $\frac{32}{2x^2} + 1 = \frac{8}{x} \rightarrow x = 4$

4) $\frac{1}{x-5} = \frac{4}{x-5} + 1 \rightarrow x = 2$

5) $\frac{1}{x^2} + \frac{1}{x} = \frac{1}{4x^2} \rightarrow x = -\frac{3}{4}$

6) $\frac{1}{10x^2} = \frac{1}{2x} + \frac{11}{10x^2} \rightarrow x = -2$

7) $\frac{1}{x^2-x} + \frac{1}{x} = \frac{10}{x^2-x} \rightarrow x = 10$

8) $\frac{x-5}{6x} = \frac{1}{5x} + 1 \rightarrow x = -\frac{31}{25}$

9) $\frac{x-1}{x} + \frac{1}{x^2+2x} = 1 , x = -1$

10) $\frac{x-2}{x+3} - 1 = \frac{3}{x+2} , x = -\frac{19}{8}$

11) $\frac{x}{10} = \frac{6}{x-4} \rightarrow x = 10, x = -6$

12) $\frac{2}{x-2} = \frac{5}{x-1} \rightarrow x = \frac{8}{3}$

13) $\frac{3}{5x} = \frac{5}{9x-2} \rightarrow x = 3$

14) $\frac{2}{2x-3} = \frac{10}{6x+1} \rightarrow x = 4$

15) $\frac{2}{x+2} + \frac{3}{x} = \frac{x}{x+2} \rightarrow x = 6, x = -1$

16) $\frac{3}{x+1} = \frac{2}{x-3} \rightarrow x = 11$

17) $\frac{x}{x-2} + \frac{1}{5} = \frac{x+1}{x-2} \rightarrow x = 7$

18) $\frac{1}{x-4} + \frac{x}{x-2} = \frac{2}{x^2-6x+8} \rightarrow x = -1$

19) $\frac{x}{x+3} = \frac{8}{x+6} \rightarrow x = 6, x = -4$

20) $\frac{x}{x-1} - \frac{2}{x} = \frac{1}{x-1} \rightarrow x = 2$

Trigonometric Functions

✍ **Evaluate.**

1) $\sin 90° = $ _____

2) $\sin -330° = $ _____

3) $\tan -30° = $ _____

4) $\sin -60° = $ _____

5) $\sin 150° = $ _____

6) $\cos 315° = $ _____

7) $\cos 180° = $ _____

8) $\cot 90° = $ _____

9) $\tan 165° = $ _____

10) $\sec 45° = $ _____

11) $\cos - 90° = $ _____

12) $\sec 60° = $ _____

13) $\csc 480° = $ _____

14) $\cot -135° = $ _____

15) $\cot 150° = $ _____

16) $\sec 120° = $ _____

17) $\csc -360° = $ _____

18) $\cot -270° = $ _____

✍ **Find the exact value of each trigonometric function. Some may be undefined.**

19) $\cot \frac{2\pi}{3} = $ _____

20) $\tan \frac{\pi}{3} = $ _____

21) $\sin \frac{2\pi}{6} = $ _____

22) $\cot \frac{5\pi}{3} = $ _____

23) $\cos - \frac{3\pi}{4} = $ _____

24) $\sec \frac{\pi}{3} = $ _____

25) $\csc \frac{5\pi}{6} = $ _____

26) $\cot \frac{4\pi}{3} = $ _____

Trigonometric Functions- Answers

✎ **Evaluate.**

1) $sin\ 90° = 1$

2) $sin -330° = \frac{1}{2}$

3) $tan -30° = -\frac{\sqrt{3}}{3}$

4) $sin -60° = -\frac{\sqrt{3}}{2}$

5) $sin\ 150° = \frac{1}{2}$

6) $cos\ 315° = \frac{\sqrt{2}}{3}$

7) $cos\ 180° = -1$

8) $cot\ 90° = 0$

9) $tan\ 165° = \sqrt{3} - 2$

10) $sec\ 45° = \sqrt{2}$

11) $cos -90° = 0$

12) $sec\ 60° = 2$

13) $csc\ 480° = \frac{2\sqrt{3}}{3}$

14) $cot -135° = -1$

15) $cot\ 165° = -2 - \sqrt{3}$

16) $sec\ 120° = -2$

17) $csc -90° = -1$

18) $cot -270° = 0$

✎ **Find the exact value of each trigonometric function. Some may be undefined.**

19) $cot\ \frac{2\pi}{3} = -\frac{\sqrt{3}}{3}$

20) $tan\ \frac{\pi}{3} = \sqrt{3}$

21) $sin\ \frac{2\pi}{6} = \frac{\sqrt{3}}{2}$

22) $cot\ \frac{5\pi}{3} = -\frac{\sqrt{3}}{3}$

23) $cos -\frac{3\pi}{4} = -\frac{\sqrt{2}}{2}$

24) $sec\ \frac{\pi}{3} = 2$

25) $csc\ \frac{5\pi}{6} = 2$

26) $cot\ \frac{4\pi}{3} = \frac{\sqrt{3}}{3}$

Trig Ratios of General Angles

✎ *Evaluate.*

1) $\tan(360) = $ _____

2) $\cos(135) = $ _____

3) $\csc(45) = $ _____

4) $\tan(180) = $ _____

5) $\sin(120) = $ _____

6) $\cot(270) = $ _____

7) $\sec(-45) = $ _____

8) $\cos(-90) = $ _____

9) $\sin(330) = $ _____

10) $\sin(315) = $ _____

11) $\sec(360) = $ _____

12) $\cos(-180) = $ _____

13) $\sin(-90) = $ _____

14) $\csc(-120) = $ _____

15) $\cos(-330) = $ _____

16) $\cot(-120) = $ _____

17) $\tan(-315) = $ _____

18) $\cot(360) = $ _____

19) $\sin(-135) = $ _____

20) $\sec(-300) = $ _____

Trig Ratios of General Angles - Answers

✎ *Evaluate.*

1) $\tan(360) = 0$

2) $\cos(135) = -\frac{\sqrt{2}}{2}$

3) $\csc(45) = \sqrt{2}$

4) $\tan(180) = 0$

5) $\sin(120) = \frac{\sqrt{3}}{2}$

6) $\cot(270) = 0$

7) $\sec(-45) = \sqrt{2}$

8) $\cos(-90) = 0$

9) $\sin(330) = -\frac{1}{2}$

10) $\sin(315) = -\frac{\sqrt{2}}{2}$

11) $\sec(360) = 1$

12) $\cos(-180) = -1$

13) $\sin(-90) = -1$

14) $\csc(-120) = -\frac{2\sqrt{3}}{3}$

15) $\cos(-330) = \frac{\sqrt{3}}{2}$

16) $\cot(-120) = \frac{\sqrt{3}}{3}$

17) $\tan(-315) = 1$

18) $\cot(360) = undefined$

19) $\sin(-135) = -\frac{\sqrt{2}}{2}$

20) $\sec(-300) = 2$

Angle and Angle Measure

 **Convert each degree measure into radians.**

1) $28° =$ _____

2) $80° =$ _____

3) $75° =$ _____

4) $92° =$ _____

5) $230° =$ _____

6) $-34° =$ _____

7) $-270° =$ _____

8) $-320° =$ _____

9) $-240° =$ _____

10) $-70° =$ _____

Convert each radian measure into degrees.

11) $\frac{1}{10}\pi =$ _____

12) $-\frac{1}{5}\pi =$ _____

13) $-\frac{2}{5}\pi =$ _____

14) $-\frac{3}{8}\pi =$ _____

15) $\frac{3}{20}\pi =$ _____

16) $-\frac{9}{2}\pi =$ _____

17) $\frac{10}{3}\pi =$ _____

18) $\frac{15}{4}\pi =$ _____

19) $-\frac{2}{5}\pi =$ _____

20) $\frac{8}{5}\pi =$ _____

Angle and Angle Measure - Answers

Convert each degree measure into radians.

1) $28° = \dfrac{7}{45}\pi$

2) $80° = \dfrac{4}{9}\pi$

3) $75° = \dfrac{5}{12}\pi$

4) $92° = \dfrac{23}{45}\pi$

5) $230° = \dfrac{23}{18}\pi$

6) $-34° = -\dfrac{17}{90}\pi$

7) $-270° = -\dfrac{3}{2}\pi$

8) $-320° = -\dfrac{16}{9}\pi$

9) $-240° = -\dfrac{4}{3}\pi$

10) $-70° = -\dfrac{7}{18}\pi$

Convert each radian measure into degrees.

11) $\dfrac{1}{10}\pi = 18°$

12) $-\dfrac{1}{5}\pi = -36°$

13) $-\dfrac{2}{5}\pi = 72°$

14) $-\dfrac{3}{8}\pi = -67.5°$

15) $\dfrac{3}{20}\pi = 27°$

16) $-\dfrac{9}{2}\pi = -810°$

17) $\dfrac{10}{3}\pi = 600°$

18) $\dfrac{15}{4}\pi = 675°$

19) $-\dfrac{2}{5}\pi = -72°$

20) $\dfrac{8}{5}\pi = 288°$

Coterminal Angles and Reference Angles

✍ **Find a positive and a negative Coterminal angles.**

1) $140° =$

Positive = _____
Negative = _____

2) $-165° =$

Positive = _____
Negative = _____

3) $190° =$

Positive = _____
Negative = _____

4) $-235° =$

Positive = _____
Negative = _____

5) $-330° =$

Positive = _____
Negative = _____

6) $-420° =$

Positive = _____
Negative = _____

✍ **Find a positive and a negative Coterminal angles.**

7) $\frac{5\pi}{4} =$

Positive = _____
Negative = _____

8) $\frac{2\pi}{9} =$

Positive = _____
Negative = _____

9) $-\frac{7\pi}{9} =$

Positive = _____
Negative = _____

10) $-\frac{8\pi}{7} =$

Positive = _____
Negative = _____

Coterminal Angles and Reference Angles - Answers

✍ **Find a positive and a negative Coterminal angles.**

1) $140° =$

Positive $= 500°$

Negative $= -220°$

4) $-235° =$

Positive $= 125°$

Negative $= -595°$

2) $-165° =$

Positive $= 195°$

Negative $= -525°$

5) $-330° =$

Positive $= 30°$

Negative $= 690°$

3) $190° =$

Positive $= 550°$

Negative $= -170°$

6) $-420° =$

Positive $= 60°$

Negative $= -780°$

✍ **Find a positive and a negative Coterminal angles.**

7) $\dfrac{5\pi}{4} =$

Positive $= \dfrac{13\pi}{4}$

Negative $= -\dfrac{3\pi}{4}$

9) $-\dfrac{7\pi}{9} =$

Positive $= \dfrac{11\pi}{9}$

Negative $= -\dfrac{25\pi}{9}$

8) $\dfrac{2\pi}{9} =$

Positive $= \dfrac{20\pi}{9}$

Negative $= -\dfrac{16\pi}{9}$

10) $-\dfrac{8\pi}{7} =$

Positive $= \dfrac{6\pi}{7}$

Negative $= -\dfrac{22\pi}{7}$

Evaluating Trigonometric Function

✎ *Find the exact value of each trigonometric function.*

1) $\cos 180° = $ _____

2) $\cos -270° = $ _____

3) $\tan 225° = $ _____

4) $\sin \frac{\pi}{4} = $ _____

5) $\csc 330° = $ _____

6) $\tan -120° = $ _____

7) $\cos 390° = $ _____

8) $\csc \frac{2\pi}{3} = $ _____

9) $\csc 270° - $ _____

10) $\tan 240° = $ _____

11) $\cos -630° = $ _____

12) $\csc \frac{5\pi}{6} = $ _____

13) $\sin -\frac{2\pi}{3} = $ _____

14) $\sec \frac{4\pi}{3} = $ _____

15) $\tan \frac{7\pi}{6} = $ _____

16) $\sin \frac{15\pi}{4} = $ _____

17) $\tan \frac{11\pi}{6} = $ _____

18) $\sec -\frac{5\pi}{6} = $ _____

19) $\sec 945° = $ _____

20) $\cos 990° = $ _____

21) $\cot 315° = $

22) $\cot -\frac{10\pi}{3} = $ _____

23) $\tan -\frac{31\pi}{6} = $ _____

24) $\tan \frac{9\pi}{4} = $ _____

25) $\csc \frac{17\pi}{6} = $ _____

Evaluating Trigonometric Function - Answers

✎ *Find the exact value of each trigonometric function.*

1) $\cos 180° = -1$

2) $\cos -270° = 0$

3) $\tan 225° = 1$

4) $\sin \frac{\pi}{4} = \frac{\sqrt{2}}{2}$

5) $\csc 330° = -2$

6) $\tan -120° = \sqrt{3}$

7) $\cos 390° = \frac{\sqrt{3}}{2}$

8) $\csc \frac{2\pi}{3} = \frac{2\sqrt{3}}{3}$

9) $\csc 270° = -1$

10) $\tan 240° = \sqrt{3}$

11) $\cos -630° = 0$

12) $\csc \frac{5\pi}{6} = 2$

13) $\sin -\frac{2\pi}{3} = -\frac{\sqrt{3}}{2}$

14) $\sec \frac{4\pi}{3} = -2$

15) $\tan \frac{7\pi}{6} = \frac{\sqrt{3}}{3}$

16) $\sin \frac{15\pi}{4} = -\frac{\sqrt{2}}{2}$

17) $\tan \frac{11\pi}{6} = -\frac{\sqrt{3}}{3}$

18) $\sec -\frac{5\pi}{6} = -\frac{2\sqrt{3}}{3}$

19) $\sec 945° = -\sqrt{2}$

20) $\cos 990° = 0$

21) $\cot 315° = -1$

22) $\cot -\frac{10\pi}{3} = -\frac{\sqrt{3}}{3}$

23) $\tan -\frac{31\pi}{6} = -\frac{\sqrt{3}}{3}$

24) $\tan \frac{9\pi}{4} = 1$

25) $\csc \frac{17\pi}{6} = 2$

Missing Sides and Angles of a Right Triangle

✎ *Find the value of x.*

1) _____

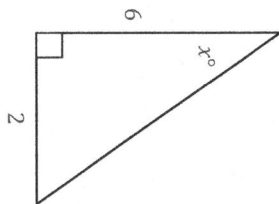

6

2

x°

2) _____

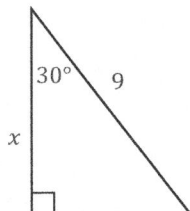

30° 9

x

3) _____

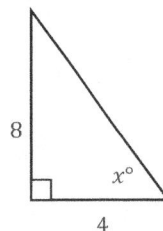

8

x°

4

4) _____

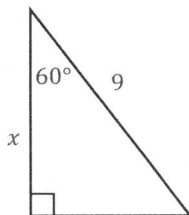

60° 9

x

5) _____

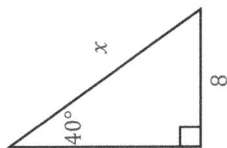

x

8

40°

6) _____

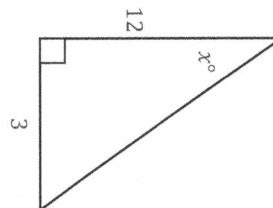

12

x°

3

7) _____

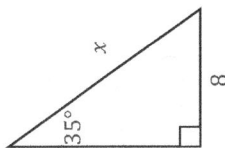

x

8

35°

8) _____

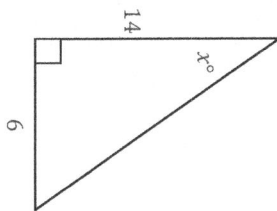

14

x°

6

9) _____

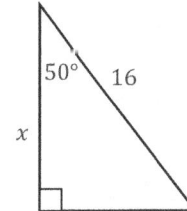

50° 16

x

10) _____

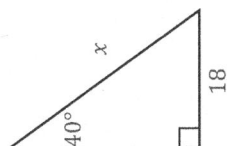

x

18

40°

11) _____

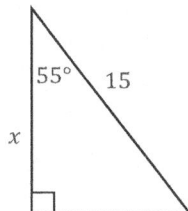

55° 15

x

12) _____

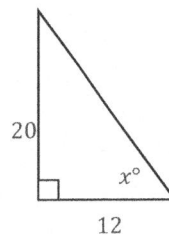

20

x°

12

Missing Sides and Angles of a Right Triangle – Answers

✎ *Find the value of* x.

1) 18

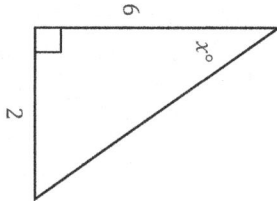

2) 7.8

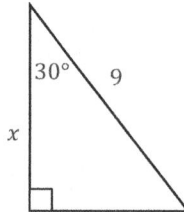

3) 63

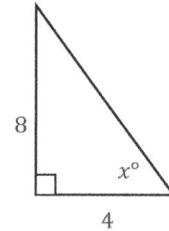

4) 4.5

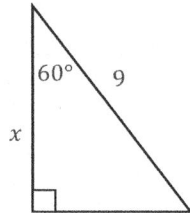

5) 12.45

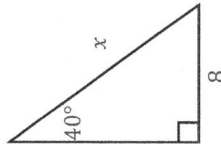

6) 14

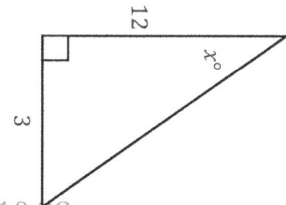

7) 13.94

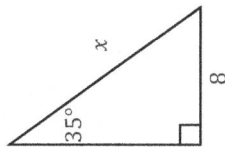

8) 23

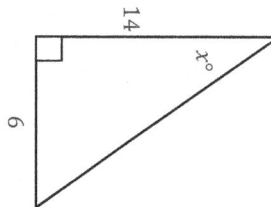

9) 10.28

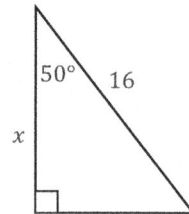

10) 28

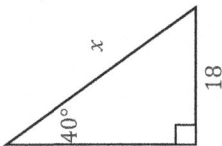

11) 8.6

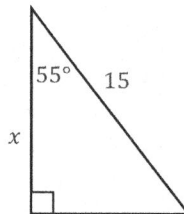

12) 59

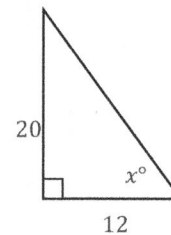

Cracking ALEKS Math Test

Embark on your journey to conquering the ALEKS Math exam!

Congratulations on reaching a vital milestone in your ALEKS Math preparation. With a solid understanding of the concepts under your belt, you're now ready to elevate your study by applying what you've learned. *"Cracking ALEKS Math Test"* is designed to bridge the gap between theory and practice, offering you a realistic testing experience.

Dive into the specifics of what this section offers:

ALEKS Math Question Types: Get acquainted with the kinds of questions to anticipate, preparing you to approach each one with a clear strategy.

How is the ALEKS Scored?: Uncover the intricacies of the scoring process to prioritize your studies and maximize your score.

ALEKS Math Test-Taking Strategies: Learn techniques that can help streamline your problem-solving process and enhance your accuracy under timed conditions.

ALEKS Math – Test Day Tips: Equip yourself with actionable advice to maintain calm and focus when it counts.

Now, put your knowledge into action:

ALEKS Math Practice Test 1 and 2: Step into the test-taker's shoes with practice tests crafted to mirror the actual exam's format and rigor.

ALEKS Math Practice Tests Answer Keys: Verify your solutions and understand your strengths and areas for improvement.

Answers and Explanations for Practice Tests 1 and 2: Benefit from detailed explanations to deepen your comprehension and correct misunderstandings.

Each practice question you work through is an opportunity to refine your skills and build confidence. Embrace this phase as a key part of your journey towards ALEKS Math success.

ALEKS Math Test-Taking Strategies

Here are some test-taking strategies that you can use to maximize your performance and results on the ALEKS Math test.

#1 : Use This Approach To Answer Every ALEKS Math Question

- Review the question to identify keywords and important information.

- Translate the keywords into math operations so you can solve the problem.

- Review the answer choices. What are the differences between answer choices?

- Draw or label a diagram if needed.

- Try to find patterns.

- Find the right method to answer the question. Use straightforward math, plug in numbers, or test the answer choices (backsolving).

- Double-check your work.

#2 : Answer Every ALEKS Math Question

Don't leave any fields empty! ALEKS is a Computer Adaptive (CA) assessment. Therefore, you cannot leave a question unanswered and you cannot go back to previous questions.

Even if you're unable to work out a problem, strive to answer it. Take a guess if you have to. You will not lose points by getting an answer wrong, though you may gain a point by getting it correct!

#3: BALLPARK

A ballpark answer is a rough approximation. When we become overwhelmed by calculations and figures, we end up making silly mistakes. A decimal that is moved by one unit can change an answer from right to wrong, regardless of the number of steps that you went through to get it. That's where ballparking can play a big part.

If you think you know what the correct answer may be (even if it's just a ballpark answer), you'll usually have the ability to estimate the range of possible answers and avoid simple mistakes.

#4: PLUGGING IN NUMBERS

"Plugging in numbers" is a strategy that can be applied to a wide range of different math problems on the ALEKS Math test. This approach is typically used to simplify a challenging question so that it is more understandable. By using the strategy carefully, you can find the answer without too much trouble.

The concept is fairly straightforward–replace unknown variables in a problem with certain values. When selecting a number, consider the following:

- Choose a number that's basic (just not too basic). Generally, you should avoid choosing 1 (or even 0). A decent choice is 2.

- Try not to choose a number that is displayed in the problem.

- Make sure you keep your numbers different if you need to choose at least two of them.

- If your question contains fractions, then a potential right answer may involve either an LCD (least common denominator) or an LCD multiple.

- 100 is the number you should choose when you are dealing with problems involving percentages.

ALEKS Mathematics – Test Day Tips

After practicing and reviewing all the math concepts you've been taught, and taking some ALEKS mathematics practice tests, you'll be prepared for test day. Consider the following tips to be extra-ready come test time.

Before Your Test

What to do the night before:

- **Relax!** One day before your test, study lightly or skip studying altogether. You shouldn't attempt to learn something new, either. There are plenty of reasons why studying the evening before a big test can work against you. Put it this way–a marathoner wouldn't go out for a sprint before the day of a big race. Mental marathoners–such as yourself–should not study for any more than one hour 24 hours before a ALEKS test. That's because your brain requires some rest to be at its best. The night before your exam, spend some time with family or friends, or read a book.

- **Avoid bright screens** - You'll have to get some good shuteye the night before your test. Bright screens (such as the ones coming from your laptop, TV, or mobile device) should be avoided altogether. Staring at such a screen will keep your brain up, making it hard to drift asleep at a reasonable hour.

- **Make sure your dinner is healthy** - The meal that you have for dinner should be nutritious. Be sure to drink plenty of water as well. Load up on your complex carbohydrates, much like a marathon runner would do. Pasta, rice, and potatoes are ideal options here, as are vegetables and protein sources.

- **Get your bag ready for test day** – Prefer to take ALEKS in the Testing Office? The night prior to your test, pack your bag with your stationery, admissions pass, ID, and any other gear that you need. Keep the bag right by your front door. If you prefer to take the test at home, find a quite place without any distractions.

- **Make plans to reach the testing site** – If you are taking the test at the testing office, ensure that you understand precisely how you will arrive at the site of the test. If parking is something you'll have to find first, plan for it. If you're dependent on public transit, then review the schedule. You should also make sure that the train/bus/subway/streetcar you use will be running. Find out about road closures as well. If a parent or friend is accompanying you, ensure that they understand what steps they have to take as well.

The Day of the Test

- Get up reasonably early, but not too early.

- **Have breakfast** - Breakfast improves your concentration, memory, and mood. As such, make sure the breakfast that you eat in the morning is healthy. The last thing you want to be is distracted by a grumbling tummy. If it's not your own stomach making those noises, another test taker close to you might be instead. Prevent discomfort or embarrassment by consuming a healthy breakfast. Bring a snack with you if you think you'll need it.

- **Follow your daily routine** - Do you watch TV in the morning while getting ready for the day? Don't break your usual habits on the day of the test. Likewise, if coffee isn't something you drink in the morning, then don't take up the habit hours before your test. Routine consistency lets you concentrate on the main objective–doing the best you can on your test.

- **Wear layers** - Dress yourself up in comfortable layers if you are taking the test at the testing site. You should be ready for any kind of internal temperature. If it gets too warm during the test, take a layer off.

- **Make your voice heard** - If something is off, speak to a proctor. If medical attention is needed or if you'll require anything, consult the proctor prior to the start of the test. Any doubts you have should be clarified. You should be entering the test site with a state of mind that is completely clear.

- **Have faith in yourself** - When you feel confident, you will be able to perform at your best. When you are waiting for the test to begin, envision yourself receiving an outstanding result. Try to see yourself as someone who knows all the answers, no matter what the questions are. A lot of athletes tend to use this technique–particularly before a big competition. Your expectations will be reflected by your performance.

During your test

- **Be calm and breathe deeply** - You need to relax before the test, and some deep breathing will go a long way to help you do that. Be confident and calm. You got this. Everybody feels a little stressed out just before an evaluation of any kind is set to begin. Learn some effective breathing exercises. Spend a minute meditating before the test starts. Filter out any negative thoughts you have. Exhibit confidence when having such thoughts.

- **Concentrate on the test** - Refrain from comparing yourself to anyone else. You shouldn't be distracted by the people near you or random noise. Concentrate exclusively on the test. If you find yourself irritated by surrounding noises, earplugs can be used to block sounds off close to you. Don't forget–the test is going to last an hour or more. Some of that time will be dedicated to brief sections. Concentrate on the specific section you are working on during a particular moment. Do not let your mind wander off to upcoming or previous questions.

- **Try to answer each question individually** - Focus only on the question you are working on. Use one of the test-taking strategies to solve the problem. If you aren't able to come up with an answer, don't get frustrated. Simply guess, then move onto the next question.

- **Don't forget to breathe!** Whenever you notice your mind wandering, your stress levels boosting, or frustration brewing, take a thirty-second break. Shut your eyes, drop your pencil, breathe deeply, and let your shoulders relax. You will end up being more productive when you allow yourself to relax for a moment.

After your test

- **Take it easy** - You will need to set some time aside to relax and decompress once the test has concluded. There is no need to stress yourself out about what you could've said, or what you may have done wrong. At this point, there's nothing you can do about it. Your energy and time would be better spent on something that will bring you happiness for the remainder of your day.

- **Redoing the test** - Did you succeed on the test? Congratulations! Your hard work paid off! Succeeding on this test means that you are now ready to take college level courses. If you didn't receive the result you expected, though, don't worry! The test can be retaken. In such cases, you will need to follow the retake policy. You also need to re-register to take the exam again.

Time to Test

Time to refine your skills with a practice test.

In this book, there are two complete ALEKS Mathematics Tests. Take these tests to the test day experience. After you've finished, score your test using the answers and explanations section.

Before You Start

- You'll need a pencil, a scientific calculator, and scratch papers to take the test.

- For these practice tests, don't time yourself. Spend time as much as you need.

- After you've finished the test, review the answer key to see where you went wrong.

Good luck!

ALEKS Math

Practice Test 1

2024

Total number of questions: 35

Total time: No time limit

Calculators are permitted for ALEKS Math Test.

1) If $f(x) = 3x + 4(x + 1) + 2$, then $f(4x) = ?$

2) Simplify $(-4 + 9i)(3 + 5i)$.

3) Simplify this expression: $\dfrac{(-2x^2 y^2)^3 (3x^3 y)}{12x^3 y^8}$.

4) Solve and write the solution set the in set-builder notation of the following inequality.

$$2x - 4(x + 2) \geq x + 6$$

5) What is the negative solution to $2x^3 - x^2 - 6x = 0$?

6) The table shows the linear relationship between the profit earned in million dollars at a home appliance factory and the number of sales of a product.

Number of Sales of a Product per thousand	1	3	7	14	15
The Profit earned in Million dollars	13	21	37	65	69

What is the rate of change in profit obtained in a million dollars with respect to the number of product sales in the factory?

7) What is the zero of $r(x) = \frac{4}{7}x + 12$?

8) The sum of $-4x^2 + 3x - 24$ and $7x^2 - 8x + 18$ can be written in the form $ax^2 + bx + c$, where a, b, and c are constants. What is the value of $a + b - c$?

9) What are the zeros of the function: $f(x) = x^2 - 7x + 12$?

10) Find the factors of the binomial $x^3 - 8$.

11) What is the value of the y-intercept of the graph of $k(x) = 42\left(\frac{4}{5}\right)^x$?

12) If $3^m \cdot 9^n = 3^{12}$, what is the value of $m + 2n$?

13) A circle is inscribed in a square and the radius of the circle is 4. What is the area of the shaded region?

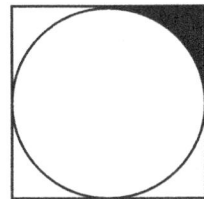

14) Given $q(x) = 3(x - 5)^2 - 7$, what is the value of $q(2)$?

15) $(7x + 2y)(5x + 2y) =$?

16) Multiply and write the product in scientific notation:

$$(2.9 \times 10^6) \times (2.6 \times 10^{-5})$$

17) Find $-2 + \frac{3b-4c}{9b} - \frac{2b+2c}{6b}$.

18) What is the number of solutions to the equation $x^2 - 3x + 1 = x - 3$?

19) Multiply and write the product in scientific notation:

$$(1.7 \times 10^9) \times (2.3 \times 10^{-7})$$

$$h = -25t^2 + st + k$$

20) The equation above gives the height h, in feet, of a ball t seconds after it is thrown straight up with an initial speed of s feet per second from a height of k feet. Find s in terms of h, t, and k.

21) Solve: $2x(5 + 3y + 2x + 4z)$.

22) What is the value of x in the following equation?

$$log_4(x + 2) - log_4(x - 2) = 1$$

23) A customer purchased movie tickets online. The total cost, c, in dollars, of t tickets can be found using the function below.

$$c = 24.50t + 5.25$$

If the customer spent a total of $103.25 on tickets, how many tickets did he purchase?

24) Mr. Anderson has a beaker containing n milliliters of a solution to distribute to the students in his chemistry class. If he gives each student 3 milliliters of the solution, he will have 5 milliliters left over. In order to give each student 4 milliliters of the solution, he will need an additional 21 milliliters. How many students are in the class?

25) Find the x −intercept and y −intercept: $3x - 6y = 24$.

26) Approximately, what is the perimeter of the figure below? ($\pi = 3$)

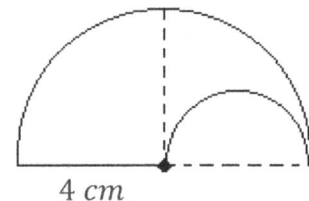

4 cm

27) What is the inverse of the function $f(x) = x^2 + 1$?

28) Perform the operations and simplify $\sqrt{8} - \sqrt{50} + \sqrt{72}$.

$$y = x^2 - 9x + 18$$

29) The equation above represents a parabola in the xy-plane. What is one x-intercept of the parabola?

30) What is the solution to the system of equations below?

$$4x - 7y = -2$$
$$12x - 21y = -42$$

31) What is the domain of $f(x) = -4x^2 + 25$?

32) The exponential growth $f(x) = 5(3)^x$ is shown in the following table. Find the average rate of change over the given interval $1 \le x \le 3$.

x	0	1	2	3	4
$f(x)$	5	15	45	135	450

33) Find $\sum_{i=1}^{n}(i + 2)$.

34) According to the graph, what is the minimum
number of degrees of the function $f(x)$?

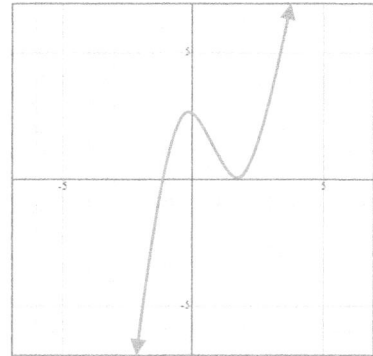

35) The table is given below for the continuous function $f(x)$. What is the minimum
degree for the function?

x	$f(x)$
-3	8
-1	2
0	-1
2	14
5	-7
10	-3
18	1

This is the end of Practice Test 1.

ALEKS Math

Practice Test 2

2024

Total number of questions: 35

Total time: No time limit

Calculators are permitted for ALEKS Math Test.

1) At a rate of $3d + 9$ Kilometers per hour, how many kilometers can a train travel in 8 hours?

2) The regular price for a concert ticket is $100. A different ticket vendor offers a 10% discount on the regular price. What would be the savings in dollars and cents if you purchase 3 tickets from the discounted vendor instead of the regular vendor?

3) Solve: $\frac{1.4 \times 10^{-7}}{2 \times 10^{-10}}$.

4) If y varies directly with x, the relationship can be represented by the equation $y = kx$, where k is the proportionality constant. Given that $y = 6$ when $x = 24$, what is the equation of the direct variation that represents this relationship?

5) If $x = 3\left(log_2 \frac{1}{32}\right)$, what is the value of x?

6) The table represents different values of function $g(x)$. What is the value of the expression $3g(-2) - 2g(3)$?

x	$g(x)$
-2	3
-1	2
0	1
1	0
2	-1
3	-2

7) If a quadratic function with equation $y = ax^2 + 5x + 10$, where a is constant, passes through the point $(2, 12)$, what is the value of a^2?

8) The graph of quadratic function f is shown on the grid.
   If $g(x) = x^2$ and $f(x) = g(x) + k$, what is the value of k?

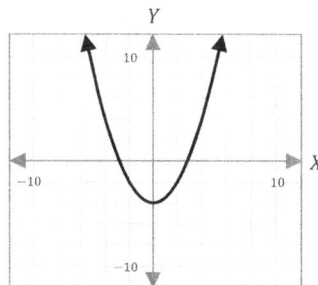

$$3x + x + x - 2 = x + x + x + 8$$

9) In the equation above, what is the value of x?

10) The area of the following equilateral triangle with sides of length d is ...

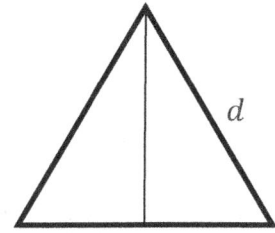

11) What is the solution to the following inequality? $|x - 9| \leq 5$

12) Express in scientific notation: 0.00095.

13) If $x^2 + \frac{1}{x^2} = 7$, find the value of $x^4 + \frac{1}{x^4}$.

14) Find a factor of $6x^2 - 4x - 10$.

15) The expression $(xy^{-2})^3 \left(\frac{y}{x}\right)^9$ is equivalent to $x^n y^m$. What is the value of $n - m$?

16) Solve: $4x - 3y < 2y + 35$.

17) For what value of x is the function $f(x)$ below undefined?

$$f(x) = \frac{1}{(x-3)^2 - 4}$$

18) What is the sum of all values of n that satisfies $2n^2 + 16n + 24 = 0$?

19) A construction worker can complete building a brick wall in 5 hours and a wooden fence in 3 hours. The function below can be used to find the number of brick walls the worker builds when she completes f wooden fences in a 40-hour workweek.

$$b = \frac{(50 - 3f)}{0.5}$$

If the worker built 10 brick walls in one week, how many wooden fences did she complete that week?

20) Subtract $5x^2 - 3$ from triple the quantity $-x^2 - 2x + 2$.

21) Simplify $(3x - 5)^2$.

22) Solve: $f(x) = 2(2 - 3x)^2 - 7$.

23) In the following figure, point O is the center of the circle and the equilateral triangle has a perimeter of 45. What is the circumference of the circle? ($\pi = 3$)

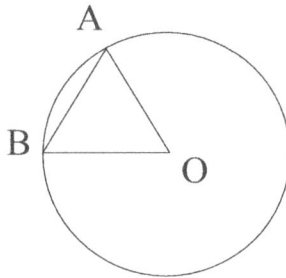

24) $(5x - 3y)^2$.

25) Solve this equation for x: $e^{3x} = 18$.

26) Simplify $\frac{4-3i}{-4i}$.

27) What is the solution to $2(n - 1) = 3(n + 2) - 10$?

28) If a and b are solutions of the following equation, what is the ratio $\frac{a}{b}$? $(a > b)$

$$2x^2 - 11x + 8 = -3x + 18$$

29) Point A lies on the line with equation $y - 3 = 2(x + 5)$. If the $x-$coordinate of A is 8, what is the $y-$coordinate of A?

30) The graph of $y = -3x^2 + 12x + 6$ is shown below. If the graph crosses the $y-$axis at the point $(0, r)$, what is the value of r?

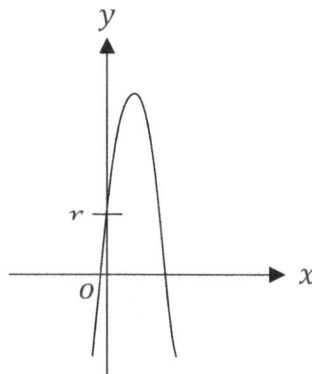

31) What is the range of the function graphed on the grid?

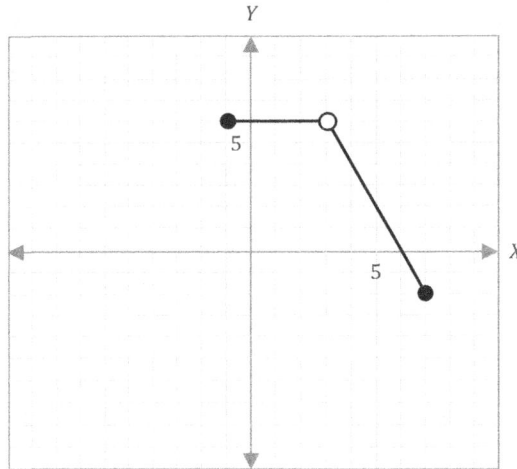

32) What expression is equivalent to $-49x^2 + 9$?

33) If $f(x) = 3x + 4(x + 1) + 2$, then $f(3x) = ?$

34) If $f(x) = \frac{10x - 3}{6}$ and $f^{-1}(x)$ is the inverse of $f(x)$, what is the value of $f^{-1}(2)$?

35) The perimeter of a rectangular garden is 38 meters. The length of the garden can be represented by $(x + 6)$ meters, and its width can be represented by $(2x - 2)$ meters. What are the dimensions of this garden in meters?

This is the end of Practice Test 2.

ALEKS Mathematics Practice Tests
Answers and Explanations

Now, it's time to review your results to see where you went wrong and what areas you need to improve!

ALEKS Mathematics Practice Test 1

1) **The answer is $28x + 6$.**

If $f(x) = 3x + 4(x + 1) + 2$, then find $f(4x)$ by substituting $4x$ for every x in the function. This gives: $f(4x) = 3(4x) + 4(4x + 1) + 2$.

It simplifies to: $f(4x) = 3(4x) + 4(4x + 1) + 2 = 12x + 16x + 4 + 2 = 28x + 6$.

2) **The answer is $-57 + 7i$.**

We know that: $i = \sqrt{-1} \Rightarrow i^2 = -1$,

$(-4 + 9i)(3 + 5i) = -12 - 20i + 27i + 45i^2 = -12 + 7i - 45 = -57 + 7i$.

3) **The answer is $\dfrac{-2x^6}{y}$.**

Using rules of exponents, start in the numerator with $(-2x^2y^2)^3$, which is $(-2)^3 (x^2)^3(y^2)^3$, which simplifies to $-8x^6y^6$. That is multiplied by $3x^3y$, giving $-24x^9y^7$.

4) Next, divide $\frac{-24x^9y^7}{12x^3y^8}$ to get $\frac{-2x^6}{y}$. **The answer is $\{x | x \leq -\dfrac{14}{3}\}$.**

Start by using the distributive property to simplify the left side of the inequality and combine like terms to get $-2x - 8 \geq x + 6$. To isolate x, subtract x and add 8 to both sides. This gives $-3x \geq 14$. To isolate the x, divide both sides by -3. Dividing by the negative changes the relationship between the sides and gives $x \leq -\dfrac{14}{3}$. In set-builder notation, the solution is: $\{x | x \leq -\dfrac{14}{3}\}$

5) **The answer is $-\dfrac{3}{2}$.**

Simplify the equation $Q(x) = 2x^3 - x^2 - 6x$. First, factor the equation: $2x^3 - x^2 - 6x = x(2x^2 - x - 6)$. To find the zeros, each factor should be equal to zero: $x(2x^2 - x - 6) = 0$. Therefore, the zeros are $x = 0$, $2x^2 - x - 6 = 0$.

At this point, evaluate the discriminant of the quadratic equation $2x^2 - x - 6 = 0$.

The expression $\Delta = b^2 - 4ac$ is called the discriminant for the standard form of the quadratic equation of the form $ax^2 + bx + c = 0$. So, $\Delta = b^2 - 4ac \rightarrow \Delta = (-1)^2 - 4(2)(-6) = 49$.

Since $\Delta > 0$, the quadratic equation has two distinct solutions. Now, we will use the quadratic formula: $x_{1,2} = \frac{-b \pm \sqrt{\Delta}}{2a}$

Then, the roots are $x_1 = \frac{-(-1)+\sqrt{49}}{2(2)} = \frac{1+7}{4} = 2$, and $x_2 = \frac{-(-1)-\sqrt{49}}{2(2)} = \frac{1-7}{4} = -\frac{3}{2}$.

6) The answer is 4.

Since there is a linear relationship between the data. So, the rate of change in this model is the same value between all points. By using the rate of change formula $\frac{f(b)-f(a)}{b-a}$, for every a and b such that $a < b$. We can evaluate the rate of change for points 1 and 3.

Therefore, $\frac{21-13}{3-1} = \frac{8}{2} = 4$.

7) The answer is −21.

To find the answer to the problem, solve the equation for $r(x) = 0$: $\frac{4}{7}x + 12 = 0$.

Subtract 12 from both sides: $\frac{4}{7}x = -12$. Multiply both sides by $\frac{7}{4}$: $x = -21$.

8) The answer is 4.

The sum of the given expressions is $(-4x^2 + 3x - 24) + (7x^2 - 8x + 18)$. Combining like terms yields $3x^2 - 5x - 6$. Based on the form of the given equation, $a = 3$, $b = -5$, and $c = -6$. Therefore, $a + b - c = 3 + (-5) - (-6) = 4$.

9) The answer is 3 and 4.

First, factor the function: $(x - 4)(x - 3)$. To find the zeros, $f(x)$ should be zero: $f(x) = (x - 4)(x - 3) = 0$.

Therefore, the zeros are, $(x - 4) = 0 \Rightarrow x = 4$, $(x - 3) = 0 \Rightarrow x = 3$.

10) The answer is $(x - 2)(x^2 + 2x + 4)$.

The formula for factoring the difference between two cubes is:

$(a^3 - b^3) = (a - b)(a^2 + ab + b^2)$. Here, we have $x^3 - 8 = x^3 - 2^3$. Replacing a with x and b with 2 gives: $(x - 2)(x^2 + 2x + 4)$

11) The answer is 42.

In order to find the y−intercept, you must put the value of the variable x in the equation of the function $k(x)$ equal to zero. Therefore, when $x = 0$, the y−intercept is obtained.

$x = 0 \rightarrow k(0) = 42 \left(\frac{4}{5}\right)^0 = 42$.

12) The answer is 12.

Change 9^n into a base 3 exponential. Since $9 = 3^2$, you can substitute 3^2 for 9.

$3^m . (3^2)^n = 3^{12} \rightarrow 3^m . 3^{2n} = 3^{12} \rightarrow 3^{m+2n} = 3^{12} \rightarrow m + 2n = 12$.

13) The answer is $16 - 4\pi$.

Subtract the area of the circle from the area of the square and divide the result by 4. The side of the square is equal to the diameter of the circle, so each side is equal to 8.

Area of the square $\rightarrow 8 \times 8 = 64$. Area of the circle $\rightarrow A = \pi r^2 = (4 \times 4)\pi = 16\pi$

Area of the shaded region $\rightarrow \frac{64 - 16\pi}{4} = \frac{64}{4} - \frac{16\pi}{4} = 16 - 4\pi$

14) The answer is 20.

Put $x = 2$ in the equation:

$q(2) = 3(2 - 5)^2 - 7 = 3(-3)^2 - 7 = 3(9) - 7 = 20$.

15) The answer is $35x^2 + 24xy + 4y^2$.

Use the FOIL (First, Out, In, Last) method:

$(7x + 2y)(5x + 2y) = 35x^2 + 14xy + 10xy + 4y^2$

$= 35x^2 + 24xy + 4y^2$

16) The answer is 7.54×10^1.

$(2.9 \times 10^6) \times (2.6 \times 10^{-5}) = (2.9 \times 2.6) \times (10^6 \times 10^{-5}) = 7.54 \times \left(10^{6+(-5)}\right)$

$= 7.54 \times 10^1$

17) The answer is $-2 - \frac{7c}{9b}$.

To write this expression as a single fraction, we need to find a common denominator. The common denominator of $9b$ and $6b$ is $18b$. Then:

$$-2 + \frac{3b-4c}{9b} - \frac{2b+2c}{6b} = \frac{-2(18b)}{18b} + \frac{2(3b-4c)}{18b} - \frac{3(2b+2c)}{18b}.$$

Now, simplify the numerators and combine them:

$$\frac{-2(18b)}{18b} + \frac{2(3b-4c)}{18b} - \frac{3(2b+2c)}{18b} = \frac{-36b}{18b} + \frac{6b-8c}{18b} - \frac{6b+6c}{18b} = \frac{-36b+6b-8c-6b-6c}{18b} = \frac{-36b-14c}{18b}.$$

Divide both the numerator and denominator by 2. Then: $\frac{-36b-14c}{18b} = \frac{-18b-7c}{9b} = -2 - \frac{7c}{9b}$.

18)   The answer is 1.

Rewrite the equation $x^2 - 3x + 1 = x - 3$ by simplifying. First, subtract x from both sides: $x^2 - 3x + 1 - x = x - 3 - x \rightarrow x^2 - 4x + 1 = -3$. Then, add 3 to both sides of the equation: $x^2 - 4x + 1 + 3 = -3 + 3 \rightarrow x^2 - 4x + 4 = 0$.

Now, remember that there can be 0, 1, or 2 solutions to a quadratic equation. In standard form, a quadratic equation is written as $ax^2 + bx + c = 0$.

For the quadratic equation, the expression $\Delta = b^2 - 4ac$ is called the discriminant. If the discriminant is positive, there are 2 distinct solutions for the quadratic equation. If the discriminant is 0, there is one solution for the quadratic equation and if it is negative the equation does not have any solutions.

Find the value of the discriminant: $\Delta = b^2 - 4ac \rightarrow \Delta = (-4)^2 - 4(1)(4) = 16 - 16 = 0$.

Since the discriminant is zero, the quadratic equation has one distinct solution.

19) The answer is 3.91×10^2.

To multiply two numbers in scientific notation, multiply their coefficients and add their exponents. For these two numbers in scientific notation, multiply the coefficients: $1.7 \times 2.3 = 3.91$. Add the powers of 10: $10^9 \times 10^{-7} = 10^{9+(-7)} = 10^2$. Then:

$$(1.7 \times 10^9) \times (2.3 \times 10^{-7}) = (1.7 \times 2.3) \times (10^9 \times 10^{-7}) = 3.91 \times \left(10^{9+(-7)}\right)$$

$$= 3.91 \times 10^2$$

20) The answer is $s = \frac{h-k}{t} + 25t$.

Starting with the original equation, $h = -25t^2 + st + k$, in order to get s in terms of the other variables, $-25t^2$ and k need to be subtracted from each side. This yields $st = h + 25t^2 - k$, which when divided by t will give s in terms of the other variables. The equation $s = \frac{h+25t^2-k}{t}$, can be further simplified. Another way to write the previous equations is $s = \frac{h-k}{t} + \frac{25t^2}{t}$, which can be simplified to $s = \frac{h-k}{t} + 25t$.

21) The answer is $10x + 6xy + 4x^2 + 8xz$.

Use the distributive property: $2x(5 + 3y + 2x + 4z) = 10x + 6xy + 4x^2 + 8xz$.

22) The answer is $\frac{10}{3}$.

We know that: $\log_a b - \log_a c = \log_a \frac{b}{c}$ and $\log_a b = c \rightarrow b = a^c$. Then:

$\log_4(x + 2) - \log_4(x - 2) = 1 \rightarrow \log_4 \frac{x+2}{x-2} = 1 \rightarrow \frac{x+2}{x-2} = 4^1 = 4 \rightarrow x + 2 = 4(x - 2)$.

Therefore: $x + 2 = 4x - 8 \rightarrow 4x - x = 8 + 2 \rightarrow 3x = 10 \rightarrow x = \frac{10}{3}$.

23) The answer is 4.

To find the number of tickets the customer purchased, we can use the given function and solve for t: $c = 24.50t + 5.25$ (Substitute $c = \$103.25$) $\rightarrow 103.25 = 24.50t + 5.25$ (Subtract 5.25 from both sides) $98 = 24.5t$

(Divide both sides by 24.5) $t = \frac{98}{24.5} = 4$. Therefore, the customer purchased 4 tickets.

24) The answer is 26.

Let c be the number of students in Mr. Anderson's class. The conditions described in the question can be represented by the equations $n = 3c + 5$ and $n + 21 = 4c$. Substituting $3c + 5$ for n in the second equation gives $3c + 5 + 21 = 4c$, which can be solved to find $c = 26$.

25) The answer is -4.

To find the $x-$intercept, put $x = 0$ in the equation $3x - 6y = 24$, then we get:

$x = 0 \rightarrow 3(0) - 6y = 24 \rightarrow -6y = 24 \rightarrow y = -4$.

So, the $y-$intercept is -4.

26) The answer is 22.

If there is a small semicircle inside a big semicircle, then: the perimeter of the big semicircle $= \frac{8\times\pi}{2} = \frac{24}{2} = 12\,cm$. The perimeter of the small semicircle $= \frac{4\times\pi}{2} = \frac{12}{2} = 6\,cm$.

Total perimeter $= 4 + 6 + 12 = 22\,cm$.

27) The answer is $f^{-1}(x) = \pm\sqrt{x-1}$.

First, replace $f(x)$ with y: $y = x^2 + 1$. Then, replace all x's with y and all y's with x: $x = y^2 + 1$.

Now, solve for y: $x = y^2 + 1 \rightarrow x - 1 = y^2 \rightarrow |y| = \sqrt{x-1} \rightarrow y = \pm\sqrt{x-1}$. Finally, replace y with $f^{-1}(x)$: $f^{-1}(x) = \pm\sqrt{x-1}$.

28) The answer is $3\sqrt{2}$.

Find the factor of the numbers:

$8 = 4 \times 2 = 2^2 \times 2$

$50 = 25 \times 2 = 5^2 \times 2$

$72 = 36 \times 2 = 6^2 \times 2$

Now use the radical rule: $\sqrt[n]{a^n} = a$.

Finally: $\sqrt{8} - \sqrt{50} + \sqrt{72} = \sqrt{2^2 \times 2} - \sqrt{5^2 \times 2} + \sqrt{6^2 \times 2} = 2\sqrt{2} - 5\sqrt{2} + 6\sqrt{2} = 3\sqrt{2}$.

29) The answer is 3 or 6.

The x−intercepts of the parabola represented by $y = x^2 - 9x + 18$ in the xy−plane are the values of x for which y is equal to 0. The factored form of the equation, $y = (x-3)(x-6)$, shows that y equals 0 if and only if $x = 3$ or $x = 6$. Thus, the x−intercepts of the parabola are 3 and 6.

30) There is no solution.

To solve this system of equations, we can use the method of elimination. If we multiply the first equation by -3, we get $-12x + 21y = 6$. If we add this equation to the second equation, we get $12x - 21y + (-12x + 21y) = -42 + (6) \rightarrow 0 = -36$.

This equation simplifies to $0 = -36$, which is not true. Therefore, there is no solution to this system of equations.

31) The answer is $(-\infty, +\infty)$.

Since the equation of the function f is quadratic, the domain of this function is all real numbers. It means that the interval $(-\infty, +\infty)$ and is represented by $\mathbb{R}$.

32) The answer is 60.

To calculate the average rate of change in the exponential growth $f(x)$, considering that the given interval $1 \leq x \leq 3$, put $a = 1$ and $b = 3$ with the corresponding values $f(a) = 15$ and $f(b) = 135$. Now, use this formula $\frac{f(b)-f(a)}{b-a}$, and substituting the values:

The average rate of change $= \frac{135-15}{3-1} = \frac{120}{2} = 60$.

33) The answer is $\frac{n(n+5)}{2}$.

Use the formula $\sum_{i=1}^{n}(a_i + b_i) = \sum_{i=1}^{n} a_i + \sum_{i=1}^{n} b_i$, so:

$\sum_{i=1}^{n}(i + 2) = \sum_{i=1}^{n} i + \sum_{i=1}^{n} 2$.

Now, using these properties $\sum_{i=1}^{n} i = \frac{n(n+1)}{2}$, and $\sum_{i=1}^{n} c = nc$. Therefore,

$\sum_{i=1}^{n}(i + 2) = \frac{n(n+1)}{2} + 2n = \frac{n^2+n+4n}{2} = \frac{n^2+5n}{2} = \frac{n(n+5)}{2}$.

34) The answer is three.

According to the given graph, the value of the function is zero for two inputs. It means that the degree of the function is greater than 2. On the other hand, on the left $f(x)$ goes to $-\infty$, and on the right $f(x)$ goes to $+\infty$, so the function of odd degree. Therefore, the degree of the function $f(x)$ is three.

35) The answer is 4.

Since the function is continuous, the function has at least one root every time it changes sign. According to the table, in the intervals of $(-1,0)$, $(0,2)$, $(2,5)$, and $(10,18)$, the values of the function change from positive to negative and vice versa. Therefore, the function $f(x)$ has at least four zero values. So, the minimum degree is 4.

ALEKS Mathematics Practice Test 2

1) The answer is $24d + 72$.

Set a proportion: $\frac{1}{8} = \frac{3d+9}{x} \rightarrow x = 8(3d + 9) = 24d + 72$.

2) The answer is $30.

If the regular price for a concert ticket is $100, and a different vendor offers a 10% discount on the regular price, then the discounted price would be:

Discounted price = Regular price − (Discount rate × Regular price)

Discounted price = $100 − (0.1 × $100) → Discounted price = $90

Therefore, each ticket costs $90 from the discounted vendor.

To calculate the savings, we need to find the difference between the total cost of purchasing 3 tickets at the regular price and the total cost of purchasing 3 tickets at the discounted price:

Regular price for 3 tickets = $100 × 3 = $300

Discounted price for 3 tickets = $90 × 3 = $270

Savings = Regular price for 3 tickets − Discounted price for 3 tickets

Savings = $300 − $270 Savings = $30

So, the savings in dollars and cents by purchasing 3 tickets from the discounted vendor instead of the regular vendor would be $30.

3) The answer is 7×10^2.

$\frac{1.4 \times 10^{-7}}{2 \times 10^{-10}} = \frac{1.4}{2} \times 10^{10-7} = 0.7 \times 10^3 = (7 \times 10^{-1}) \times 10^3 = 7 \times 10^{3-1} = 7 \times 10^2$

4) The answer is $y = 0.25x$.

To find the equation of the relationship of direct variation, it is enough to find the value of k. Substitute the given values of x and y into the formula $y = kx$ and evaluate the value of k. So, we get:

$\begin{matrix} x = 24 \\ y = 6 \end{matrix} \rightarrow 6 = 24k \rightarrow k = \frac{6}{24} = k = \frac{1}{4}$ or 0.25.

Next, put the obtained value in the formula: $y = 0.25x$.

5) The answer is −15.

We know $log_a \frac{x}{y} = log_a x - log_a y$. Therefore:

$x = 3\left(log_2 \frac{1}{32}\right) \to x = 3(log_2 1 - log_2 32)$. In addition, we know $log_a 1 = 0$.

Therefore: $x = -3 log_2 32$. Considering: $log_a x^b = b \times log_a x$, and $log_a a = 1$.

Then: $log_2 32 = log_2 2^5 = 5 log_2 2 = 5$. Finally, we have:

$x = -3 log_2 32 \to x = -3 \times 5 \to x = -15$.

6) The answer is 13.

Based on the table provided: $g(-2) = g(x = -2) = 3 \to g(3) = g(x = 3) = -2$

$3g(-2) - 2g(3) = 3(3) - 2(-2) = 9 + 4 = 13$.

7) The answer is 4.

Plug in the values of x and y of the point $(2, 12)$ in the equation of the quadratic function. Then:

$12 = a(2)^2 + 5(2) + 10 \to 12 = 4a + 10 + 10 \to 12 = 4a + 20$

$\to 4a = 12 - 20 = -8 \to a = \frac{-8}{4} = -2 \to a^2 = (-2)^2 = 4$

8) The answer is −2.

Notice that for the function $g(x)$, if $k > 0$ then $f(x) = g(x) + k$ is shifted up by $|k|$ units, and if $k < 0$ the function $f(x) = g(x) + k$ is shifted down by $|k|$ units. The graph of the function $g(x) = x^2$ is as follows:

If $g(x)$ is shifted down by 2 units, the graph of $f(x)$ is obtained. Therefore, the value of k is -2 and the corresponding equation is $f(x) = g(x) - 2 \to f(x) = x^2 - 2$.

9) The answer is 5.

$3x + x + x - 2 = x + x + x + 8$, combining like terms on each side of the given equation yields $5x - 2 = 3x + 8$. Adding 2 to both sides of the equation $5x - 2 + 2 = 3x + 8 + 2 \to 5x = 3x + 10$.

Subtracting $3x$ from both sides gives $5x - 3x = 3x + 10 - 3x \to 2x = 10$.

Divide both sides of $2x = 10$ by 2 to yield $x = 5$.

10) The answer is $\frac{\sqrt{3}\,d^2}{4}$.

You can also find the height of the triangle using the relationship in $30° - 60° - 90°$ triangles. The relationship among all sides of the special right triangle $30° - 60° - 90°$ is provided in this triangle:

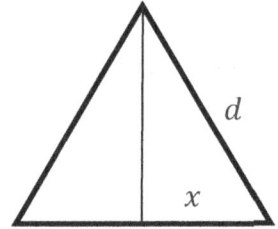

In the equilateral triangle, the side is d. Then, $2x = d \rightarrow x = \frac{d}{2}$

The height of the triangle is $\frac{d}{2} \times \sqrt{3} = \frac{d\sqrt{3}}{2}$. Then, the area of the triangle is:

$\frac{1}{2}(d)\left(\frac{d\sqrt{3}}{2}\right) = \frac{\sqrt{3}\,d^2}{4}$.

The area of the equilateral triangle is $\frac{\sqrt{3}\,d^2}{4}$.

11) The answer is $4 \le x \le 14$.

Since this inequality contains an absolute value, then, the value inside the absolute value bars is greater than -5 and less than 5. Then:

$|x - 9| \le 5 \rightarrow -5 \le x - 9 \le 5 \rightarrow -5 + 9 \le x - 9 + 9 \le 5 + 9 \rightarrow 4 \le x \le 14$.

12) The answer is 9.5×10^{-4}.

The number 0.00095 is equal to 9.5×10^{-4} in scientific notation.

13) The answer is 47.

First, using $(x + y)^2 = x^2 + 2xy + y^2$. Substitute x by x^2, and y by $\frac{1}{x^2}$. We have:

$\left(x^2 + \frac{1}{x^2}\right)^2 = x^4 + \frac{1}{x^4} + 2x^2 \times \frac{1}{x^2} \rightarrow \left(x^2 + \frac{1}{x^2}\right)^2 = x^4 + \frac{1}{x^4} + 2$.

Substitute $x^2 + \frac{1}{x^2} = 7$. Therefore:

$(7)^2 = x^4 + \frac{1}{x^4} + 2 \rightarrow 49 = x^4 + \frac{1}{x^4} + 2 \rightarrow x^4 + \frac{1}{x^4} = 47$.

14) The answer is $2(x + 1)(3x - 5)$.

$6x^2 - 4x - 10 = 2(x + 1)(3x - 5)$

15) The answer is -9.

For the expression $(xy^{-2})^3$, use the exponential rule as $(xy)^a = x^a \times y^a$. Then:

$(xy^{-2})^3 = x^3 \times (y^{-2})^3$

In this case, by using the rule $(x^a)^b = x^{a \times b}$, we have: $(y^{-2})^3 = y^{-2 \times 3} = y^{-6}$. Substitute the expression in the resulting expression from $(xy^{-2})^3$:

$(xy^{-2})^3 = x^3 y^{-6}$

Considering the rule: $\left(\frac{a}{b}\right)^c = \frac{a^c}{b^c}$ for the expression $\left(\frac{y}{x}\right)^9$, then: $\left(\frac{y}{x}\right)^9 = \frac{y^9}{x^9} = y^9 x^{-9}$.

Substitute $y^9 x^{-9}$ and $x^3 y^{-6}$ in the composite expression: $(xy^{-2})^3 \left(\frac{y}{x}\right)^9 = x^3 y^{-6} y^9 x^{-9}$.

Arrange the terms to have the same base in the expression $x^3 y^{-6} y^9 x^{-9}$ to form $x^3 x^{-9} y^{-6} y^9$. According to the exponential rule: $x^a \times x^b = x^{a+b}$. Thus, $x^3 x^{-9} y^{-6} y^9 = x^{3-9} y^{-6+9} = x^{-6} y^3$.

Put $x^n y^m = x^{-6} y^3$. The values n and m are -6 and 3, respectively. Therefore,

$n - m = -6 - (3) = -9$.

16) The answer is $y > 0.8x - 7$.

To find the inequality equivalent to $4x - 3y < 2y + 35$, we can start by isolating y on one side of the inequality:

$4x - 3y - 2y < 2y + 35 - 2y \rightarrow 4x - 5y < 35 \rightarrow 4x - 5y - 4x < 35 - 4x$

$\rightarrow -5y < 35 - 4x$

Divide both sides of the inequality by -5. So, the equivalent inequality is $y > \frac{4}{5}x - \frac{35}{5}$, which means that the answer is $y > 0.8x - 7$.

17) The answer is 1 or 5.

The function $f(x)$ is undefined when the denominator of $\frac{1}{(x-3)^2 - 4}$ is equal to zero. Find the values of x for which the equation $(x-3)^2 - 4 = 0$ holds. We get:

$(x-3)^2 - 4 = 0 \rightarrow (x-3)^2 = 4 \rightarrow |x - 3| = 2$

So, $x - 3 = 2 \rightarrow x = 5$, or $x - 3 = -2 \rightarrow x = 1$.

Therefore, the value of x for which $f(x)$ is undefined is 1 or 5.

18) The answer is −8.

The problem asks for the sum of the roots of the quadratic equation $2n^2 + 16n + 24 = 0$. Dividing each side of the equation by 2 gives $n^2 + 8n + 12 = 0$. If the roots of $n^2 + 8n + 12 = 0$ are n_1 and n_2, then the equation can be factored as $n^2 + 8n + 12 = (n - n_1)(n - n_2) = 0$. Looking at the coefficient of n on each side of $n^2 + 8n + 12 = (n + 6)(n + 2)$ gives $n = -6$ or $n = -2$, then, $-6 + (-2) = -8$.

19) The answer is 15.

To solve the problem, plug the given information into the equation and solve for the variable b:
$$10 = \frac{(50 - 3f)}{0.5}$$
Multiplying both sides by 0.5: $5 = 50 - 3f$. Subtracting 50 from both sides: $-45 = -3f$. Dividing both sides by -3: $f = 15$. So, the construction worker completes 15 wooden fences that week.

20) The answer is $-8x^2 - 6x + 9$.

$3(-x^2 - 2x + 2) - (5x^2 - 3) = -3x^2 - 6x + 6 - 5x^2 + 3 = -8x^2 - 6x + 9$

21) The answer is $9x^2 - 30x + 25$.

Use the FOIL (First-Out-In-Last) method to simplify the expression:

$(3x - 5)^2 = (3x - 5)(3x - 5) = 9x^2 - 15x - 15x + 25 = 9x^2 - 30x + 25$.

22) The answer is $18x^2 - 24x + 1$.

Use the polynomial identity: $(x - y)^2 = x^2 - 2xy + y^2$ for the part of the equation of function $(2 - 3x)^2$. Then:

$(2 - 3x)^2 = (2)^2 - 2(2)(3x) + (3x)^2 = 4 - 12x + 9x^2$.

Substitute the obtained expression in the equation of function and simplify.

$2(2 - 3x)^2 - 7 = 2(4 - 12x + 9x^2) - 7 = 18x^2 - 24x + 1$.

Therefore: $f(x) = 18x^2 - 24x + 1$.

23) The answer is 90.

In the equilateral triangle if x is the length of one side of the triangle, then the perimeter of the triangle is $3x$. Then $3x = 45 \rightarrow x = 15$ and the radius of the circle is $x = 15$. Then, the circumference of the circle is:

$2\pi r = 2\pi(15) = 30\pi$, $\pi = 3 \rightarrow 30\pi = 30 \times 3 = 90$.

24) The answer is $25x^2 + 9y^2 - 30xy$.

Perfect square formula: $(a - b)^2 = a^2 - 2ab + b^2$. So, $(5x - 3y)^2 =$

$25x^2 - 30xy + 9y^2 = 25x^2 + 9y^2 - 30xy$

25) The answer is $\frac{ln(18)}{3}$.

If $f(x) = g(x)$, then: $ln(f(x)) = ln(g(x)) \rightarrow ln(e^{3x}) = ln(18)$.

Use the logarithm rule: $log_a x^b = b \, log_a x \rightarrow ln(e^{3x}) = 3x \, ln(e) \rightarrow (3x) \, ln(e) = ln(18)$

$ln(e) = 1$, then: $(3x) ln(e) = ln(18) \rightarrow 3x = ln(18) \rightarrow x = \frac{ln(18)}{3}$

26) The answer is $\frac{3}{4} + i$.

Recall that the imaginary numbers (containing i) cannot be in the denominator of a fraction. To simplify the expression, multiply both the numerator and denominator by i.

$\frac{4-3i}{-4i} \times \frac{i}{i} = \frac{4i-3i^2}{-4i^2}$, $i^2 = -1$, then: $\frac{4i-3i^2}{-4i^2} = \frac{4i-3(-1)}{-4(-1)} = \frac{4i+3}{4} = \frac{4i}{4} + \frac{3}{4} = \frac{3}{4} + i$

27) The answer is $n = 2$.

Simplify and solve for n in the equation.

$2(n - 1) = 3(n + 2) - 10 \rightarrow 2n - 2 = 3n + 6 - 10 \rightarrow 2n - 2 = 3n - 4$

Subtract $2n$ from both sides: $-2 = n - 4$, add 4 to both sides: $n = 2$.

28) The answer is -5.

$2x^2 - 11x + 8 = -3x + 18 \rightarrow 2x^2 - 11x + 3x + 8 - 18 = 0 \rightarrow 2x^2 - 8x - 10 = 0$

$\rightarrow 2(x^2 - 4x - 5) = 0 \rightarrow$ Divide both sides by 2. Then: $x^2 - 4x - 5 = 0$, find the factors of the quadratic equation. $\rightarrow (x - 5)(x + 1) = 0 \rightarrow x = 5$ or $x = -1$. $a > b$, then: $a = 5$ and $b = -1$.

$\frac{a}{b} = \frac{5}{-1} = -5$

29) The answer is 29.

Here we can substitute 8 for x in the equation. Thus, $y - 3 = 2(8 + 5)$, $y - 3 = 26$.

Adding 3 to both sides of the equation: $y = 26 + 3$, $y = 29$.

30) The answer is 6.

Since the graph crosses the $y-$axis at $(0, r)$, then substituting 0 for x and r for y in

$r = -3(0)^2 + 12(0) + 6$ creates a true statement: $r = -3(0)^2 + 12(0) + 6$, or $r = 6$.

31) The answer is $-2 \leq y \leq 6$.

The range of the function is the possible value for y. The image of the graph on the

$y-$axis is equivalent to the range of the graph. Look at the following graph:

The interval $-2 \leq y \leq 6$ is the range of the function.

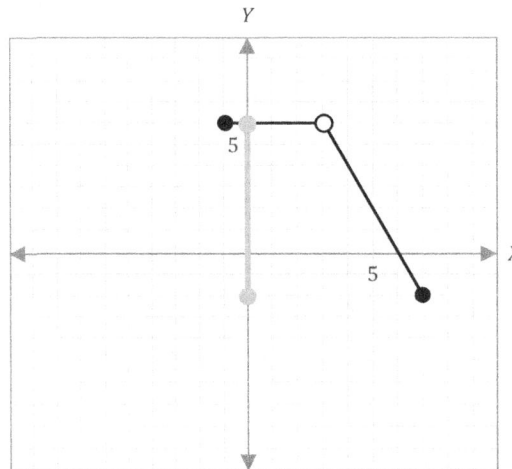

32) The answer is $(3 + 7x)(3 - 7x)$.

Rewrite the expression as $9 - 49x^2$. Use the following polynomial identity:

$x^2 - y^2 = (x + y)(x - y)$.

Then: $9 - 49x^2 = (3 + 7x)(3 - 7x)$.

33) The answer is $21x + 6$.

If $f(x) = 3x + 4(x + 1) + 2$, then find $f(3x)$ by substituting $3x$ for every x in the

function. This gives: $f(3x) = 3(3x) + 4(3x + 1) + 2$

It simplifies to: $f(3x) = 3(3x) + 4(3x + 1) + 2 = 9x + 12x + 4 + 2 = 21x + 6$.

34) The answer is $\frac{3}{2}$.

To solve for the inverse function, first, replace $f(x)$ with y. Then, solve the equation for x and after that, replace every x with y and replace every y with x. Finally, replace y with $f^{-1}(x)$.

$$f(x) = \frac{10x - 3}{6} \Rightarrow y = \frac{10x - 3}{6} \Rightarrow 6y = 10x - 3 \Rightarrow 6y + 3 = 10x \Rightarrow \frac{6y + 3}{10} = x$$

$$f^{-1}(x) = \frac{6x + 3}{10} \Rightarrow f^{-1}(2) = \frac{6(2) + 3}{10} = \frac{15}{10} = \frac{3}{2}$$

35) The answer is $11\,m$ by $8\,m$.

To solve the problem, we can use the formula for the perimeter of a rectangle, which is: Perimeter $= 2 \times$ Length $+ 2 \times$ Width.

We are given that the perimeter of the rectangle is 38 centimeters. We are also given that the length can be represented as $(x + 6)$ and the width can be represented as $(2x - 2)$. So, we can substitute these values into the formula for the perimeter and solve for x:

$$38 = 2(x + 6) + 2(2x - 2) \rightarrow 38 = 2x + 12 + 4x - 4 \rightarrow 38 = 6x + 8$$

$$\rightarrow 30 = 6x \rightarrow x = 5.$$

Now, we can find the length and width of the rectangle by substituting $x = 5$ into the expressions for the length and width:

Length $= x + 6 = 5 + 6 = 11\,m$, and width $= 2x - 2 = 2(5) - 2 = 8\,m$.

Therefore, the dimensions of the rectangle are $11\,m$ by $8\,m$.

... So Much More Online!

Effortless Math Online ALEKS Math Center offers a complete study program, including the following:

✓ Step-by-step instructions on how to prepare for the ALEKS Math test

✓ Numerous ALEKS Math worksheets to help you measure your math skills

✓ Complete list of ALEKS Math formulas

✓ Video lessons for ALEKS Math topics

✓ Full-length ALEKS Math practice tests

✓ And much more...

No Registration Required.

Visit **EffortlessMath.com/ALEKS** to find your online ALEKS Math resources.

The Best ALEKS Math Books!
Download eBooks (in PDF format) Instantly!

Most Popular ALEKS Math Books!

ALEKS MATH Practice Workbook 2024

The Most Comprehensive Review for the Math Section of the ALEKS TEST

2 full-length ALEKS Math practice tests

Test Taker's #1 Choice

Comprehensive ALEKS MATH WorkBook

Visit EffortlessMath.com/ALEKS for Online ALEKS Math Resources

Recommended by Test Prep Experts

Reza Nazari

Download at

Download

The Ultimate Step by Step Guide to Preparing for the ALEKS Math Test

Test taker's #1 Choice

ALEKS Math 2024 for Beginners

Reza Nazari

Recommended by Test Prep Experts

Download at

Download

ALEKS Math in 10 Days

The Most Effective ALEKS Math Crash Course

Test Taker's #1 Choice

This Book Will Help You

Visit www.EffortlessMath.com for Online Math Practice

Reza Nazari

Recommended by Test Prep Experts

Download at

Download

Everything You Need to Help Achieve an Excellent Score

ALEKS Math Full Study Guide

Comprehensive Review + Practice Tests + Online Resources

2024 2025

Recommended by Test Prep Experts

Reza Nazari

Download at

Download

Receive the PDF version of this book or get another FREE book!

Thank you for using our Book!

Do you LOVE this book?

Then, you can get the PDF version of this book or another book absolutely FREE!

Please email us at:

info@EffortlessMath.com

for details.

Author's Final Note

I hope you enjoyed reading this book. You've made it through the book! Great job!

First of all, thank you for purchasing this practice book. I know you could have picked any number of books to help you prepare for your ALEKS Math test, but you picked this book and for that I am extremely grateful.

It took me years to write this practice book for the ALEKS Math because I wanted to prepare a comprehensive ALEKS Math book to help test takers make the most effective use of their valuable time while preparing for the test.

After teaching and tutoring math courses for over a decade, I've gathered my personal notes and lessons to develop this practice test. It is my greatest hope that the practice tests in this book could help you prepare for your test successfully.

If you have any questions, please contact me at reza@effortlessmath.com and I will be glad to assist. Your feedback will help me to greatly improve the quality of my books in the future and make this book even better. Furthermore, I expect that I have made a few minor errors somewhere in this book. If you think this to be the case, please let me know so I can fix the issue as soon as possible.

If you enjoyed this book and found some benefit in reading this, I'd like to hear from you and hope that you could take a quick minute to post a review on the book's Amazon page.

I personally go over every single review, to make sure my books really are reaching out and helping students and test takers. Please help me help ALEKS Math test takers, by leaving a review!

I wish you all the best in your future success!

Reza Nazari

Math teacher and author